KT-442-253

Regulation of Gene Expression by Hormones

BIOCHEMICAL ENDOCRINOLOGY

Series Editor: Kenneth W. McKerns

STRUCTURE AND FUNCTION OF THE GONADOTROPINS
Edited by Kenneth W. McKerns

SYNTHESIS AND RELEASE OF ADENOHYPOPHYSEAL
HORMONES
Edited by Marian Jutisz and Kenneth W. McKerns

REPRODUCTIVE PROCESSES AND CONTRACEPTION
Edited by KennethW. McKerns

HORMONALLY ACTIVE BRAIN PEPTIDES: Structure and Function
Edited by Kenneth W. McKerns and Vladimir Pantić

REGULATION OF GENE EXPRESSION BY HORMONES
Edited by Kenneth W. McKerns

Regulation of Gene Expression by Hormones

Edited by

Kenneth W. McKerns

International Foundation for Biochemical Endocrinology
Blue Hill Falls, Maine

PLENUM PRESS • *NEW YORK AND LONDON*

Library of Congress Cataloging in Publication Data

Main entry under title:

Regulation of gene expression by hormones.

(Biochemical endocrinology)
Sponsored by the International Foundation for Biochemical Endocrinology.
Bibliography: p.
Includes index.
Contents: Hormone action in human breast cancer/David J. Adams, Dean P.
Edwards, and William L. McGuire—Modulation of an estrogen-induced protein in the
MCF$_7$ human mammary cancer cells/Henri Rochefort—The estrogen-induced/depen-
dent renal adenocarcinoma of the Syrian hamster/Paul H. Naylor...[et al.]—[etc.]
  1. Gene expression—Addresses, essays, lectures. 2. Genetic regulation—Addresses,
essays, lectures. 3. Hormones—Addresses, essays, lectures. I. McKerns, Kenneth W. II.
International Foundation for Biochemical Endocrinology. III. Series. [DNLM: 1. Gene
expression regulations. 2. Hormones—Physiology. WK 102 R3435]
QH450.R42 1983   599′.087′322   83-6310
ISBN 0-306-41204-7

BROMLEY COLLEGE OF TECHNOLOGY		
ACCN.		~~9~~ _B 09120_
CLASSN.	574.87	
CAT.	LOCN.	

©1983 Plenum Press, New York
A Division of Plenum Publishing Corporation
233 Spring Street, New York, N.Y. 10013

All rights reserved

No part of this book may be reproduced, stored in a retrieval system, or transmitted
in any form or by any means, electronic, mechanical, photocopying, microfilming,
recording, or otherwise, without written permission from the Publisher

Printed in the United States of America

Contributors

David J. Adams, Department of Medicine, University of Texas Health Science Center, San Antonio, Texas 78284

Jutta Arnemann, Institut für Physiologische Chemie der Phillips-Universität, 3550 Marburg, G.F.R.

Miguel Beato, Institut für Physiologische Chemie der Phillips-Universität, 3550 Marburg, G.F.R.

Robert T. Chatterton, Jr., Department of Obstetrics and Gynecology, Northwestern University Medical School, Chicago, Illinois 60611

Claude DeLouis, Laboratoire de Physiologie de la Lactation, Institut National de la Recherche Agronomique, C.N.R.Z., 78350 Jouy-en-Josas, France

Eve Devinoy, Laboratoire de Physiologie de la Lactation, Institut National de la Recherche Agronomique, C.N.R.Z., 78350 Jouy-en-Josas, France

Jean Djiane, Laboratoire de Physiologie de la Lactation, Institut National de la Recherche Agronomique, C.N.R.Z., 78350 Jouy-en-Josas, France

Dean P. Edwards, Department of Medicine, University of Texas Health Science Center, San Antonio, Texas 78284

John N. Haan, Department of Obstetrics and Gynecology, Northwestern University Medical School, Chicago, Illinois 60611

Louis-Marie Houdebine, Laboratoire de Physiologie de la Lactation, Institut National de la Recherche Agronomique, C.N.R.Z., 78350 Jouy-en-Josas, France

Sheila M. Judge, Department of Obstetrics and Gynecology, Northwestern University Medical School, Chicago, Illinois 60611

Junzo Kato, Department of Obstetrics and Gynecology, Yamanashi Medical University, Tamaho, Nakakoma-gun, Yamanashi Prefecture, 409-38 Japan

Alvin M. Kaye, Department of Hormone Research, Weizmann Institute of Science, Rehovot 76100, Israel

Paul A. Kelly, Département d'Endocrinologie Moléculaire, Centre Hospitalier de l'Université Laval, Quebec, Canada

John T. Knowler, Department of Biochemistry, University of Glasgow, Glasgow G12 8QQ, Scotland

Rabinder N. Kurl, Department of Biological Chemistry and Laboratory of Human Reproduction and Reproductive Biology, Harvard Medical School, Boston, Massachusetts 02115

Janet M. Loring, Department of Biological Chemistry and Laboratory of Human Reproduction and Reproductive Biology, Harvard Medical School, Boston, Massachusetts 02115

William L. McGuire, Department of Medicine, University of Texas Health Science Center, San Antonio, Texas 78284

Synthia H. Mellon, Metabolic Research Unit, Department of Medicine, University of California, San Francisco, California 94143

Carla Menne, Institut für Physiologische Chemie der Phillips-Universität, 3550 Marburg, G.F.R.

Walter L. Miller, Department of Pediatrics, and Metabolic Research Unit, Department of Medicine, University of California, San Francisco, California 94143

Heidrun Müller, Institut für Physiologische Chemie der Phillips-Universität, 3550 Marburg, G.F.R.

Paul H. Naylor, Department of Biological Chemistry and Laboratory of Human Reproduction and Reproductive Biology, Harvard Medical School, Boston, Massachusetts 02115

David Olive, Department of Obstetrics and Gynecology, Northwestern University Medical School, Chicago, Illinois 60611

Michèle Ollivier-Bosquet, Laboratoire de Physiologie de la Lactation, Institut National de la Recherche Agronomique, C.N.R.Z., 78350 Jouy-en-Josas, France

Tsuneko Onouchi, Department of Obstetrics and Gynecology, Teikyo University School of Medicine, Kaga, Itabashi, Tokyo, 173 Japan

Malcolm G. Parker, Imperial Cancer Research Fund, P.O. Box 123, Lincoln's Inn Fields, London WC2A 3PX, U.K.

Dietmar Richter, Institut für Physiologische Chemie, Abteilung Zellbiochemie, Universität Hamburg, D-2000 Hamburg 20, G.F.R.

Henri Rochefort, Unité d'Endocrinologie Cellulaire et Moléculaire, U 148 INSERM, 34100 Montpellier, France

Hartwig Schmale, Institut für Physiologische Chemie, Abteilung Zellbiochemie, Universität Hamburg, D-2000 Hamburg 20, G.F.R.

Eldon D. Schriock, Department of Obstetrics and Gynecology, Northwestern University Medical School, Chicago, Illinois 60611

Jean-Luc Servely, Laboratoire de Physiologie de la Lactation, Institut National de la Recherche Agronomique, C.N.R.Z., 78350 Jouy-en-Josas, France

Guntram Suske, Institut für Physiologische Chemie der Phillips-Universität, 3550 Marburg, G.F.R.

Bertrand Teyssot, Laboratoire de Physiologie de la Lactation, Institut National de la Recherche Agronomique, C.N.R.Z., 78350 Jouy-en-Josas, France

Harold G. Verhage, Department of Obstetrics and Gynecology, University of Illinois at the Medical Center, Chicago, Illinois 60680

Claude A. Villee, Department of Biological Chemistry and Laboratory of Human Reproduction and Reproductive Biology, Harvard Medical School, Boston, Massachusetts 02115

Michael Wenz, Institut für Physiologische Chemie der Phillips-Universität, 3550 Marburg, G.F.R.

Preface

The International Foundation for Biochemical Endocrinology is incorporated as a nonprofit research and educational organization. It is dedicated to the dissemination of knowledge, cooperative research programs, and cultural interaction on an international basis. The Foundation is concerned with both basic research and practical applications of biological knowledge to the betterment of humanity. Among our interests are global resource management, human reproduction, hormonal regulation of normal and cancer cells, study of aging and degenerative diseases, brain peptides, peptide neurotransmitter compounds, mechanism of action of hormones, peptide hormone synthesis, and recombinant DNA techniques.

This monograph is the ninth sponsored by the Foundation in the *Biochemical Endocrinology* series. The previous four have been: *Hormonally Active Brain Peptides: Structure and Function* (1982), K. W. McKerns and V. Pantić, eds.; *Reproductive Processes and Contraception* (1981), K. W. McKerns, ed.; *Synthesis and Release of Adenohypophyseal Hormones* (1980), M. Jutisz and K. W. McKerns, eds.; and *Structure and Function of the Gonadotropins* (1978), K. W. McKerns, ed. These have all been published by Plenum Press.

This monograph deals with some novel and interesting aspects of gene regulation, including the regulation of normal and neoplastic eukaryotic cells by hormones in model systems. The search is for the molecular train of events induced. The studies follow several approaches, such as RNA–DNA hybridization surveys of the newly synthesized mRNA as well as detailed investigations of hormone-induced proteins from specialized cells. Also, by recombinant DNA technology, the expression and organization of genes was studied by cloning DNA molecules specific for mRNAs for use as DNA probes and for isolation of DNA clones.

Several chapters discuss how hormones, estrogens in particular, induce and regulate tumor growth in hormone-dependent cancer cells. An interesting variant to *in vivo* animal studies is provided by a chapter on the use of human mammary cancer cells to study estrogen-induced protein. These chapters are followed by chapters on the hormone regulation of the synthesis of milk fat in mammary epithelial cells and on the expression of prolactin and casein genes in the mammary cell.

The mechanism of estrogen action is again a dominant theme in chapters on sequential regulation of gene expression and on uterine hypertrophy. The emphasis here is on estrogen-induced protein and estrogen stimulation, mRNA, rRNA, and tRNA synthesis, as well as the nature of the proteins on which increased production of ribosomes might depend. These include RNA polymerase, ribosomal proteins, ribonuclease, and nonhistone chromatin proteins.

There follows a chapter on the molecular mechanisms by which eukaryotic cells regulate the expression of specific genes. The model here is the regulation of the uteroglobin gene. This detailed chapter includes the structure and function of uteroglobin, its distribution and hormonal control, preuteroglobin mRNA, hormonal regulation of transcription of its gene, cloning and characterization of preuteroglobin cDNA, and, finally, isolation and structural analysis of the uteroglobin gene. Another chapter deals with the evolution and regulation of genes for growth hormone and prolactin. Of interest is the use of cultured rat pituitary cells and transcription of nuclei *in vitro,* as well as the use of many techniques of recombinant DNA technology.

The next chapter, on androgenic control of gene expression in prostate, involves cloning of prostate cDNAs, expression of RNA with isolation of prostatic-binding protein genes, and their characterization.

The final chapters deal with estrogen and progesterone receptors in the brain and regulation of peptide hormone synthesis in the hypothalamus at the transcriptional level, as well as posttranscriptional processing and modification.

There was no meeting based on the monograph *Regulation of Gene Expression by Hormones.* A meeting of the Foundation in late September, 1982, was held in Geilo, Norway, to present and discuss chapters to be included in the monograph, *Regulatory Mechanisms in Target-Cell Responsiveness.* Professor Asbjørn Aakvaag and Dr. Vidar Hanssen arranged the marvelous hotel and conference facilities as well as the social and cultural activities. This was also the opportunity to discuss new and exciting research developments, to plan the future activities of the Foundation, and to arrange research programs between our members' laboratories.

Kenneth W. McKerns

Blue Hill Falls, Maine

Contents

8 Regulation of the Expression of the Uteroglobin Gene by Ovarian Hormones

Miguel Beato, Jutta Arnemann, Carla Menne, Heidrun Müller, Guntram Suske, and Michael Wenz

9 Evolution and Regulation of Genes for Growth Hormone and Prolactin

Walter L. Miller and Synthia H. Mellon

10 Androgenic Control of Gene Expression in Rat Ventral Prostate

Malcolm G. Parker

Hormone Action in Human Breast Cancer
Estrogen Regulation of Specific Proteins

David J. Adams, Dean P. Edwards, and
William L. McGuire

1. Introduction

How do estrogens regulate mammary tumor growth and function? Researchers have been pondering this question ever since breast tumor regression was reported in premenopausal patients following oophorectomy (Beatson, 1896). However, systematic exploitation of endocrine ablative surgery did not occur until the 1950s when Huggins and Dao (1954) obtained 38 clinical remissions after performing 100 ovariectomies for palliation of advanced breast cancer. A similar result occurred with adrenalectomy (Huggins and Bergenstal, 1952) and patients responding to this surgery were found to concentrate [^{3}H]hexestrol in their metastases (Folca et al., 1961). Jensen et al. (1967) and Terenius (1968) then observed that the clinical response to endocrine therapy was correlated with specific binding of estradiol by certain human breast tumor biopsy specimens. This observation and the discovery of a high-affinity receptor protein for estradiol in the cytoplasm of target cells (Toft and Gorski, 1966) led to the widespread use of an estrogen receptor (ER) assay to identify hormone-responsive breast tumors (McGuire et al., 1975). Subsequent studies have shown that approximately two-thirds of all breast tumor biopsy specimens contain estrogen receptors, and, of these tumors,

David J. Adams, Dean P. Edwards, and William L. McGuire • Department of Medicine, University of Texas Health Science Center, San Antonio, Texas 78284.

that about half respond to ablative and additive endocrine therapies (Edwards *et al.*, 1979). Furthermore, ER analysis of the primary tumor can predict response to endocrine therapy if inaccessible metastatic disease develops. Receptor-rich primary tumors display a lower probability and rapidity of recurrence (Knight *et al.*, 1977; Jensen, 1981), while absence of ER in the primary tumor is prognostic of a higher rate of recurrence and shorter survival (Osborne *et al.*, 1980). Loss of estrogen receptor may in fact represent a critical stage in breast tumor progression, since ER-negative (ER −) tumors seem to be less differentiated and more aggressive. In addition, tumors lacking ER tend to metastasize to visceral organs such as brain and liver while ER-positive (ER +) tumors are more likely to spread to bone (Singhakowinta *et al.*, 1976; Stewart *et al.*, 1981). These clinical findings have further intensified efforts to understand at the molecular level why some breast tumors exhibit estrogen-dependent growth while other tumors do not, despite retention of estrogen receptor activity, and to dissect the molecular events that occur during the uncoupling and eventual loss of the estrogen receptor pathway.

The development of breast tumor cell lines that retain most, if not all, of the receptors for steroid hormones (Horwitz *et al.*, 1978) has provided a convenient model system in which to study hormone action in breast cancer. We now know, for example, that estrogen can stimulate the synthesis of specific proteins in cultured breast tumor cells (described later), which is a recognized mode of hormone action. However, it is also apparent that estradiol *fails* to evoke responses in cell cultures that are observed *in vivo*. Perhaps the most perplexing observation is that estrogens display little, if any, mitogenic activity toward breast tumor cells *in vitro*. This paradox probably results from our incomplete knowledge of culture requirements for hormone-regulated growth. Indeed, only recently have we come to realize the importance of hormonal effects at the cell surface, particularly membrane–substratum interactions. On the other hand, there is a distinct possibility that estrogens may not act directly on breast tumor cells at all (Sirbasku and Benson, 1980; Shafie, 1980).

In this review we first consider normal mammary tissue as a target for estrogen action to anticipate estrogenic effects in neoplastic breast tissue. We then focus on a specific mode of hormone action: estrogen regulation of specific protein synthesis in cultured breast tumor cells. Finally, we discuss future directions in models for and concepts of hormone action in human breast cancer.

2. Are Breast Tumors True Targets for Estrogen Action?

It is generally agreed that a steroid hormone environment is required for mammary tumor growth in many species. However, a direct and pivotal role for estrogens in regulation of tumor growth remains controversial. Essentially, we wish to know if breast tumors resemble rat uterus, chick oviduct, or avian liver, the classic targets for estrogen action. These tissues possess a well-characterized estrogen receptor system and exhibit dramatic responses to estrogen either in cellular proliferation (uterus) or in production of large amounts of secretory proteins (egg proteins in oviduct and liver). Estrogen stimulation of ovalbumin synthesis in chick oviduct has been a particularly useful model for our understanding of the structure and regulation of hormone-dependent gene expression (Chambon *et al.*, 1979; McKnight and Palmiter, 1979; Chan and O'Malley, 1976). Here, estrogen appears to act exclusively by increasing transcription of the ovalbumin gene. A similar mechanism holds for induction of vitellogenin synthesis in *Xenopus laevis* and avian liver (Tata, 1979; Deeley *et al.*, 1977; Baker and Shapiro, 1977). In the rat uterus, estrogenic regulation of glucose-6-phosphate dehydrogenase occurs at both the transcriptional and translational levels. The hormone may also act indirectly by affecting the levels of $NADP^+$, the enzyme cofactor (Barker *et al.*, 1981).

Do breast tumors retain similar mechanisms for estrogen action? Before addressing this question, it might be helpful to ask if *normal* mammary epithelial tissue displays target tissue properties. Unlike uterus, oviduct, or liver, the role of estrogen in mammary development and function is much less clear, probably because hormone regulation of this tissue is so diverse and complex. To summarize briefly a recent review of this subject (Topper and Freeman, 1980), estradiol is the one of several hormones required for mammary growth during both adolescence and pregnancy, although a direct effect is uncertain. Estradiol appears to have a permissive role in differentiated breast tissue, making the epithelial cells competent to respond to other hormones. For example, estrogen alone does not induce synthesis of milk proteins *in vitro,* but is necessary for optimal expression of lactogenic enzymes in response to thyroid hormone, insulin, glucocorticoid, and prolactin (Bolander and Topper, 1979).

Thus, it is reasonable to expect that breast tumors may not exhibit classic target responses to estrogen. In fact, evidence from rat, mouse, and human breast tumors maintained in organ culture suggests that estrogens are weak mitogens at best and do not appear to cause pronounced changes in specific protein synthesis (Sirbasku and Benson, 1980). In a provocative paper, King (1979) has concluded that steroids are modulating agents in breast tumor cells and not the switch operators we have come to expect in target tissues. He argues that hormones need only be weak mitogens to produce the small change in tumor growth rate necessary to account for clinical responses to endocrine therapy. Furthermore, steroids may also exert indirect influence by affecting blood supply, cell–cell interaction, and the immune response. Sirbasku and co-workers (Sirbasku and Benson, 1980; Sirbasku, 1980) have extended this idea to propose that estrogens need not act directly on breast tumors at all, but can induce production of specific polypeptide growth factors, or estromedins, in traditional target tissues that subsequently stimulate tumor growth *in vivo*. As yet, no such estromedin has been identified, although growth factor activity has been partially purified from rat uterine, plasma, and mammary tumor extracts and is apparently associated with rat serum albumin (Sirbasku, 1980). Shafie (1980) has also found evidence for an indirect mechanism of estrogen action in breast tumors. The MCF-7 human breast cancer cell line does not require estrogen for growth in culture. Yet, when these cells are inoculated into ovariectomized nude mice, cell growth and solid-tumor formation are dependent on concomitant estrogen administration. If these tumors are then cultured *in vitro*, they revert to an estrogen-independent pattern of growth. These results are interpreted as evidence that breast tumor cell growth is normally inhibited *in vivo* and that estrogen stimulates synthesis of a gene product that blocks this inhibition. Whether this gene product comes from the host or from the tumor cell itself is unknown.

The scenario of estrogen action in normal and neoplastic breast tissue is therefore superficially different from that of classic target tissues. The hormone may act indirectly in concert with other hormones or growth factors, while direct effects may be subtle rather than dramatic. The latter point is illustrated when total cellular poly(+) RNA from 7,12-dimethyl benz(α)anthracene (DMBA)-induced rat mammary adenocarcinomas is compared to that from normal mammary glands of midpregnant rats. Molecular hybridization and cell-free translation analysis do not reveal a major class of tumor-specific sequences (Supowit and Rosen, 1980). Thus, levels of specific proteins may be changed by regulation of the relative abundancies of certain mRNA species rather than by true suppression or induction of specific genes. If induction of new gene expression does occur, only a small number of proteins representing a small fraction of total protein synthesis may be involved. For example, the glucocorticoid "domain" in hepatoma cells includes perhaps 10 proteins out of more than 1000 revealed by two-dimensional gel analysis (Ivarie and O'Farrell, 1978). Yet, subtle changes in protein synthesis may lead to profound effects. An analogy may be the estrogenic modulation of rat uterine induced protein (IP) (Notides and Gorski, 1966). This protein, which represents less than 0.1% of new protein synthesis, has been proposed as a single "key intermediary protein" through which estrogen triggers synthesis of a protein cascade, which in turn brings about the diverse uterine responses (Baulieu *et al.*, 1972). More recent evidence indicates that IP is synthesized constitutively in the rat uterus and is present in other tissues (Kaye and Reiss, 1980; Skipper *et al.*, 1980), but is induced by estrogen only in target cells. It has been shown to possess creatine kinase and enolase activities (Kaye and Reiss, 1980; Reiss and Kaye, 1981), but as yet, the precise role of IP in the overall uterotrophic response to estrogen remains a mystery. Do small amounts of estrogen-regulated IP(s) control breast tumors? The precedent for such a possibility certainly exists; indeed, an "IP-like" protein has been reported in MCF-7 cells (Mairesse *et al.*, 1980); however, its relation to uterine IP and to estrogen action in breast cancer is unknown.

3. Estrogen-Regulated Protein Synthesis in Human Breast Cancer

To understand better the mechanisms of estrogen action in mammary tumor development and to identify clinical markers for hormone-sensitive breast cancer, several laboratories have studied estrogen regulation of specific proteins in human breast tumor cell lines. This work may be roughly divided into two categories: estrogen-regulated functions or activities in which actual quantitation of the protein itself is lacking, and estrogen regulation of the amounts of specific proteins whose function is unknown.

3.1. Estrogen Regulation of Specific Biological Activities

3.1.1. Progesterone Receptor

Progesterone receptor (PgR) was the first protein shown to be regulated specifically by estradiol in human breast cancer cells (Horwitz and McGuire, 1977a). MCF-7 cells grown on medium containing calf serum (stripped of endogenous estrogen by charcoal treatment) contain a low but consistent basal level of PgR. Addition of 1 nM estradiol stimulates PgR three- to fourfold by 4 days. The response is dose-dependent and closely parallels accumulation and processing of estrogen receptor complex in the nucleus. It has been postulated that estrogen stimulation of progesterone receptor indicates presence of a functional pathway of estrogen action in breast tumor cells (Horwitz *et al.*, 1975). Clinical measurement of both ER and PgR are consistent with this proposal, identifying a subset of ER-positive patients who have response rates to endocrine therapy approaching 80% (Osborne *et al.*, 1980). Although progesterone receptor has proven to be valuable in clinical diagnosis, it is present in such small amounts and ligand binding is so labile that purification of PgR is difficult and therefore its use as a research tool is limited. Consequently, other estrogen-regulated proteins have been sought.

3.1.2. Growth-Associated Enzymes

Our laboratory has analyzed two enzymes in MCF-7 cells that may be related to tumor growth. Lactate dehydrogenase (LDH), an enzyme thought to be related to the degree of tumor malignancy (Goldman *et al.*, 1964; Hilf *et al.*, 1976) and to be involved in metabolic functions crucial to growth, has been shown to be elevated twofold by 10-nM estradiol treatment (Burke *et al.*, 1978). Curiously, only the fifth isozyme of LDH could be detected. Maximal stimulation occurs after 10 days, making the LDH response, like that of PgR, a late effect of estrogen stimulation. We have also examined DNA polymerase α activity in

MCF-7 cells (Edwards *et al.*, 1980b). This enzyme is known to increase dramatically during the S phase of the cell cycle (Lockwood *et al.*, 1967) and is therefore a logical choice as a marker for estrogen-regulated growth. Initial experiments did not reveal any estrogenic stimulation of cell growth or polymerase activity above control levels measured out to 8 days. Cell growth and enzyme activity could, however, be inhibited by growing cells on the antiestrogen nafoxidine (1 μM). If nafoxidine-pretreated cells are subsequently switched to medium containing estradiol (10 nM), cell growth and DNA polymerase activity increase fourfold after 4 days. These results suggest that MCF-7 cells may be replicating at a maximal rate in the absence of exogenous estrogen due to other growth factors in serum-containing medium. However, estrogen receptors are apparently able to regulate MCF-7 cell growth, since inhibition and subsequent "rescue" of growth and DNA polymerase activity are events specifically associated with the estrogen receptor.

Lippman and co-workers (Bronzert *et al.*, 1981) have measured another enzyme closely correlated with DNA synthesis in MCF-7 cells. Cytoplasmic thymidine kinase, an enzyme in the salvage pathway of deoxynucleotide biosynthesis, was found to increase in specific activity twofold when assayed 24 hr after estradiol addition. Stimulation of enzyme activity was dose-dependent and paralleled the dose curve for thymidine incorporation into DNA. Enzyme activity was inhibited by the antiestrogen, tamoxifen, which also inhibited cell growth. Kinetic studies on thymidine kinase from MCF-7 cytosols suggest that estrogen may act by increasing the V_{max} of the enzyme rather than the K_m, and thus indicate the presence of more active enzyme. However, care must be taken when analyzing enzyme kinetics in crude preparations. Furthermore, estrogen effects on thymidine kinase (and for that matter on PgR, LDH, and DNA polymerase) do not necessarily reflect changes in intracellular concentration of the protein, only changes in protein activity.

3.1.3. Proteolytic Enzymes

A characteristic feature of breast tumors is the ability to secrete proteolytic enzymes. These proteases may have normal roles in tissue remodeling, such as ovulation, blastocyst implantation, and in involution of mammary gland after lactation (Poole *et al.*, 1980). The fact that these enzymes are mitogens for normal cells in culture suggests that proteases may also have a role in growth control and malignant transformation (Cunningham *et al.*, 1979; Quigley *et al.*, 1980).

Studies of breast tumor explants cultured *in vitro* have shown that malignant adenocarcinomas, nonmalignant fibroadenomas, and normal breast specimens all release similar amounts of neutral proteases such as collagenase and plasminogen activator into culture media (Poole *et al.*, 1980). Similar results obtain for cathepsin D activity. However, a thiol protease has been detected selectively in carcinomas. This protease resembles cathepsin B in substrate specificity but has distinct physical characteristics. More important, estradiol can stimulate secretion of the thiol protease in certain adenocarcinoma specimens.

Secretion of proteolytic activity can also be observed in breast tumor cell lines. Hakim (1980) has reported estrogen stimulation of thiol protease activity in human mammary carcinoma cells. Furthermore, a correlation was found between estrogen stimulation of several types of protease activity and the levels of ER and PgR in the cells. Highest basal and induced levels of protease activity were found in ER + PgR + lines, while normal mammary epithelial cells (ER − PgR −) had low endogenous levels that were not affected by hormone treatment.

The MCF-7 cell line secretes elastinolytic (Hornebeck *et al.*, 1980), plasminogen activator (Butler *et al.*, 1979), and collagenase activities (Shafie and Liotta, 1980). Doses of estradiol that do not stimulate growth are nevertheless able to increase plasminogen activator activity 1.5-fold in as little as 8 hr. The response is dose-dependent (increasing to 2.6-fold) and is inhibited by tamoxifen and by inhibitors of RNA and protein synthesis. These results suggest that estrogen regulation of plasminogen activator is due to receptor-mediated increases in specific protein synthesis. One consequence of plasminogen activator synthesis may be subsequent activation of latent collagenase via conversion of the zymogen plasminogen to plasmin. This process apparently occurs in the ZR-75-1 cell line among others (Paranjpe *et al.*, 1980) and may be responsible for the protease activity against Type I (stoma and bone) and Type II (basement membrane) collagen seen in MCF-7 cells by Shafie and Liotta (1980). These authors report a two- to threefold increase in Type I collagenase activity by estradiol and a similar stimulation of both Types I and II collagenase by insulin. Since castration or diabetes prevents metastasis formation by MCF-7 cells injected into athymic nude mice, hormone regulation of collagenase activity may be involved in the metastatic potential of these cells. In fact, estrogenic stimulation of Type I collagenase may account for the ability of MCF-7 cells to erode bone *in vitro,* independent of osteoclast action (Martin *et al.*, 1980). Finally, plasminogen activation in breast tumor cytosols generates various cleavage products of estrogen receptor (Sherman *et al.*, 1980; Miller *et al.*, 1981). Observation of different species of steroid receptors due to protease action may imply a role for these enzymes in receptor activation and processing.

3.2. Estrogen-Regulated Proteins of Unknown Function

It is obvious that estrogen stimulation of human breast tumors or tumor cell lines can result in modulation of protein activities that could regulate tumor growth and development. What is needed now is to translate these effects on activities into quantitative changes in the levels of specific proteins. Such a transition usually requires purification of the protein of interest and production of a monospecific antibody for use in a quantitative immunological assay. As we have stated previously, this is often no easy task, especially if one has to deal with limited amounts of labile protein. An alternative approach to this problem is to identify estrogen-regulated proteins using some high-resolution technique for protein separation. With the advent of two- (O'Farrell, 1975) and even three- (Skipper *et al.*, 1980) dimensional gel electrophoresis systems, this approach has become fruitful.

3.2.1. The 46K Glycoprotein

Westley and Rochefort (1979) were the first to use successfully two-dimensional gel analysis of estrogen-regulated proteins in human breast cancer. They discovered a glycoprotein of 46,000 daltons (46K), pI 5.5–6.5 that is secreted into the culture medium as early as 12 hr after hormone treatment. Induction of 46K is specific for ER + breast cancer lines and is not detected in receptor-negative malignant or nonmalignant lines or in human milk or cystic disease fluid (Westley and Rochefort, 1980). The 46K protein is induced only by steroids known to interact with the estrogen receptor; similarly, the protein is repressed by the antiestrogen tamoxifen and hydroxytamoxifen which inhibit growth in MCF-7. Since these antiestrogens are partly estrogenic and can induce PgR in MCF-7 cells, induction of 46K may be more related to growth than is PgR and thus may be a better marker for estrogen-responsive breast tumors if an antibody to 46K can be raised. As with progesterone receptor, this could be a problem since intracellular levels of 46K can not be measured, possibly because it is rapidly secreted and accounts for only 0.15% of total [^{35}S]methionine incorporation into soluble protein. However, it may be that 46K exists intracellularly in a different form. For example, 46K may be leaked into the medium following estrogen-regulated changes in the cell surface. Once exposed to medium, 46K may be modified sufficiently that two-dimensional gel analysis makes the protein appear distinct from the intracellular species. This idea is consistent with data from Mairesse *et al.* (1980) who find an intracellular 46K protein stimulated by 3-hr estrogen treatment of MCF-7 cells. Although the size and charge of their protein is similar to the secretory 46K, no definite conclusion regarding the identity of the two proteins could be drawn. A more recent report from this group (Mairesse *et al.*, 1981) now indicates the intracellular 46K may be larger than first thought, and may be related to the 54K protein identified by our laboratory (described later).

3.2.2. The 24K Cytosol Protein

In our search for estrogen-regulated proteins in human breast cancer, we have chosen to use the double-label ratio method of Notides and Gorski (1966) that identified rat uterine induced protein. In this method control cells are pulse-labeled with [^{14}C]leucine, while experimental cells are labeled with [^{3}H]leucine.

The cells are then mixed, proteins extracted, and aliquots analyzed on single- or double-dimensional polyacrylamide gels. An estradiol-stimulated increase in the rate of synthesis of a specific protein relative to other cellular proteins is reflected by an increase in the $^3H/^{14}C$ ratio in a particular band or spot on the gel. Although high resolution of proteins can be obtained, this method does have some notable drawbacks. First, only mixed samples from control and experimental groups can be analyzed; thus, "nontarget" tissues are not easily assayed. Second, the method determines *relative* increases in specific proteins so that a large increase in general protein synthesis can mask individual increases. Ratio peaks can also be masked if the protein population in a limited molecular weight range is too complex. This is a particular problem if single-dimension gels are used. On the other hand, the sample must contain enough proteins with sufficiently high incorporation that a ratio baseline can be established. Finally, one must be cautious in ascribing ratio peaks to changes in protein synthesis. Protein size and charge could be modified by hormone action (e.g., estrogen-regulated proteolytic activity), as well as protein compartmentalization, either of which could conceivably cause a ratio change in the absence of actual effects on specific protein synthesis.

Our initial attempts to identify estrogen-regulated proteins in MCF-7 cells by the double-labeling technique did not reveal any prominent ($\geq$ twofold) increases in ratio. Considering our experience with estrogenic stimulation of DNA polymerase, we decided to use the nafoxidine rescue protocol, a method by which we can consistently observe estrogen stimulation of cell growth. Rescue from antiestrogen growth inhibition is specific for estrogens (Edwards *et al.*, 1980a; Zava and McGuire, 1978), and likely results from estrogen displacement of antiestrogen bound to receptor. Horwitz *et al.* (1981) found that estrogen rescue of antiestrogen-treated MCF-7 cells actually amplified the induction of progesterone receptor above levels obtained by treating with estradiol alone. The mechanism of this "superinduction" of PgR is unknown but suggests that the rescue protocol may maximize cell sensitivity to estrogen, perhaps by displacing residual endogenous estrogen and, in effect, creating a truly estrogen-withdrawn cell. This explanation is supported by evidence that even charcoal-stripped serum still retains conjugated estrogens which can be cleaved to biologically active hormones by MCF-7 cells (Vignon *et al.*, 1980). Alternatively, antiestrogen pretreatment may allow enhanced expression of certain differentiated functions simply by slowing the cellular growth rate. Whatever the exact mechanism, we reasoned that the rescue protocol would provide a consistent basal level of protein synthesis and magnify estrogen stimulation of specific proteins. Consequently, in the studies that follow, we have compared synthesis rates of specific proteins between nafoxidine-treated (control) and estrogen-"rescued" (experimental) cells. Full details of our procedures are described in previous publications (Edwards *et al.*, 1980a; Adams *et al.*, 1980; Edwards *et al.*, 1981).

Figure 1 shows a double-label analysis of cytoplasmic proteins after 6 days of estradiol stimulation. This is a single-dimension SDS gel fractionating in the

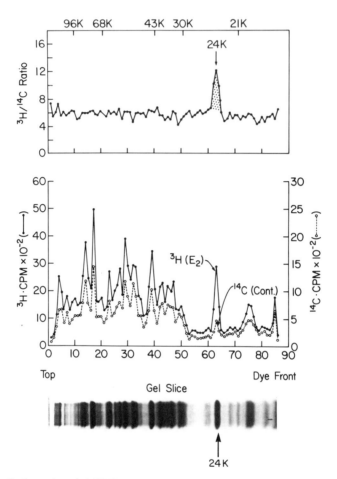

Figure 1.   Coelectrophoresis in SDS polyacrylamide of cytosol proteins synthesized by nafoxidine-treated cells ([14C]leucine) and cells "rescued" from nafoxidine inhibition by incubation with estradiol ([3H]leucine). Cells were pretreated for 6 days with 1.0 μM nafoxidine and either continued on nafoxidine for another 6 days (reference cells), or changed to medium with 10 nM estradiol for 6 days. Equal numbers of control and estrogen-stimulated cells were combined and a mixed cytosol was prepared and analyzed by SDS-polyacrylamide gel electrophoresis. The lower panel is a profile of the total 3H and 14C counts in each gel slice and the top is a plot of the corresponding 3H/14C ratios. A photograph of the Coomassie-stained gel is at the bottom of the figure. An arrow indicates the molecular weight (24K) and position of the increased 3H/14C ratio. Mobilities of molecular weight standards are indicated along the top margin.

94,000- to 14,000-mol. wt. range. The profile of the total ^{3}H and ^{14}C counts in each gel slice is indicated in the middle panel, while the upper panel gives the corresponding ^{3}H/^{14}C ratios. Under these gel conditions a single, prominent ratio peak is found at 24,000 mol. wt. coincident with a major radioactive and Coomassie-blue-staining band. We refer to this protein as 24K. Stimulation of the 24K ratio peak occurs when the isotopes are reversed and does *not* occur if cells are rescued with ethanol vehicle alone (data not shown). Thus, stimulation of 24K is due to estrogen treatment and is not due to an isotope effect.

Because we have used an antiestrogen rescue protocol, some important control experiments were necessary. We first examined the generality of the antiestrogen pretreatment. As Fig. 2 illustrates, 24K stimulation does not require nafoxidine as antiestrogen. Tamoxifen and CI-628 are equally effective. A more difficult question to address is whether increased synthesis of 24K is merely a nonspecific result of growth stimulation. Thus, the estrogen effect may be an indirect one exerted through a return to cellular proliferation and not via protein synthesis. We therefore did a serum rescue experiment to mimic nonspecific growth stimulation. After plating, MCF-7 cells were switched to serum-free

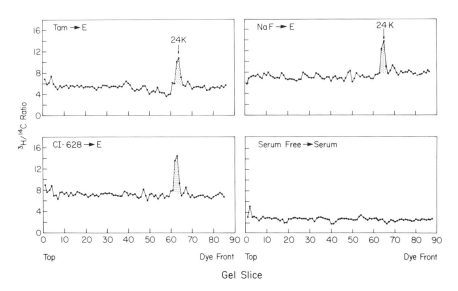

Figure 2. Comparison of specific protein synthesis between serum-stimulated and estrogen-stimulated cells. MCF-7 cells were preincubated with three different antiestrogens, tamoxifen (Tam), CI-628, and nafoxidine (Naf), followed by incubation with estradiol (→E). Double-label cytosols were prepared and analyzed by SDS electrophoresis as described in Fig. 1. Double-label cytosols prepared from serum-deprived cells (labeled with [^{14}C]leucine) and cells restimulated by addition of serum (labeled with [^{3}H]leucine) were also analyzed by SDS electrophoresis (serum-free → serum).

medium for 6 days. Under serum-free conditions, growth is arrested but cells remain viable. On day 6, control cells were continued on serum-free medium, while the experimental group was returned to medium containing serum. Although this serum rescue protocol results in a growth stimulation comparable to estrogen rescue (Edwards *et al.*, 1981), no ratio change at 24K (or anywhere else on the gel) was apparent (Fig. 2, right panels), despite the fact that *both* groups were pulse-labeled in medium containing serum. Furthermore, no relative change in 24K synthesis occurred at earlier time points of serum rescue (data not shown). Finally, 24K could represent a specific "nafoxidine-suppressed" protein. To test this possibility, cells treated with ethanol vehicle alone and pulse-labeled with [^{14}C]leucine were mixed with cells exposed to nafoxidine for 6 days and labeled with [^{3}H]leucine. Under these conditions, an antiestrogen-suppressed protein should exhibit a *negative* ratio peak, but again no ratio change at 24,000 daltons was found (Edwards *et al.*, 1981).

Confident that we were observing an estrogenic effect on specific protein synthesis, we set out to characterize the 24K response. We first tested the hormone specificity of 24K stimulation. Nafoxidine-pretreated cells were rescued with high doses of progesterone, cortisol, and a physiological dose of dihydrotestosterone (DHT). As shown in Table I, these hormones had no effect on synthesis of 24K. However, incubation with micromolar levels of DHT did stimulate 24K as might be expected, since we have previously shown that high doses of DHT elicit estrogenic responses in MCF-7 via binding to the estrogen receptor (Zava and McGuire, 1978). We then determined the estradiol dose–response for 24K stimulation and found a progressive increase in the response throughout the physiological range of hormone with maximal stimulation at 10 nM estradiol (Fig. 3).

Table I. Hormone Specificity of 24K Protein
Stimulation

Hormone	Dose	Magnitude of stimulation[a]
		% Elevation of ^{3}H/^{14}C at 24K
Estradiol 17β	10 nM	63
Progesterone	1 μM	8
Cortisol	1 μM	7
DHT[b]	10 nM	<5
DHT	1 μM	54

[a] Nafoxidine-pretreated cells were incubated for 6 days with various hormones at the indicated concentrations. Cells incubated with hormone were pulsed with [^{3}H]leucine in the same manner as estradiol-treated cells and were then mixed with ^{14}C-labeled control cells. The resulting double-label cytosols were analyzed by SDS electrophoresis as previously described.
[b] 5α-dihydrotestosterone.

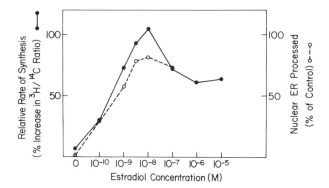

Figure 3.   Effect of different estradiol concentrations on synthesis of the 24K protein. Nafoxidine-pretreated cells were incubated for 6 days with the concentrations of estradiol indicated and double-label cytosols were prepared and analyzed by SDS electrophoresis as previously described. The degree of stimulation of the 24K protein (●—●) was estimated by calculating the percentage increase in the $^3H/^{14}C$ ratio at 24K over the baseline ratio. In a parallel experiment, cells were also preincubated with nafoxidine and then with the concentrations of estradiol indicated and 24 hr later, nuclear ER was measured by protamine sulfate exchange assay (○---○). The amount of ER processed was estimated by taking the decrease in ER content at each dosage of estradiol (compared with the level of ER in cells treated with nafoxidine only) and expressing this as a percentage of control; the control (or 100% processing level) was the processed level of nuclear ER in cells incubated for the entire period of the experiment with 10 nM estradiol.

Previous work from our laboratory with estrogen stimulation of progesterone receptor has shown that this effect is dependent on nuclear processing of ER (Horwitz *et al.*, 1978). Although nafoxidine binds and translocates ER, nuclear processing does not occur so that nuclear ER levels remain elevated. Subsequent treatment with estradiol displaces nafoxidine from receptor, processing ensues, and the PgR response occurs. We were therefore interested in knowing whether the 24K response was also related to nuclear ER processing. Figure 3 also shows that in cells treated with estradiol alone, the processed or steady state level of nuclear ER is about 25% of that in cells treated with nafoxidine. Cytosol levels of ER remain low in both groups and are less than 10% of the total. If nafoxidine-pretreated cells (containing nonprocessed nuclear ER) are subsequently incubated with increasing doses of estradiol, nuclear ER levels decrease progressively. No change in nuclear ER concentration occurs if cells are incubated with ethanol vehicle alone. Analysis of the relative rate of 24K synthesis under these conditions indicates that estrogen stimulation of 24K parallels nuclear ER processing, again suggesting that regulation of the 24K protein is a receptor-mediated event.

A central problem in detection of estrogen-regulated proteins using gel analysis is that little information is gained regarding protein identity or function. The protein can, however, be characterized by its molecular weight and charge. Figure 4 shows a two-dimensional gel of MCF-7 double-labeled cytosol after 6

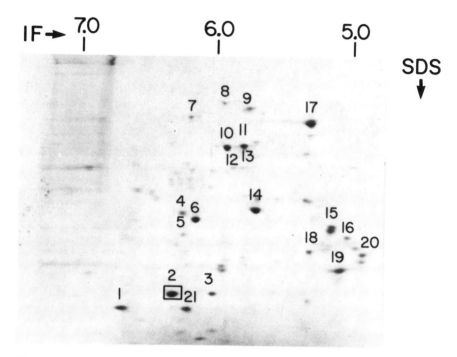

Figure 4. Two-dimensional electrophoresis of ^{3}H/^{14}C cytosols from 6 days of stimulation with 10 nM estradiol. An isoelectric focusing gel was applied directly to further separation by SDS-slab gel electrophoresis. Selected Coomassie-blue-stained spots on the second-dimension SDS gel were numbered, cut out, and counted for ^{3}H- and ^{14}C-radioactivity. Numbers on the SDS gel indicate the stained spot below it and the enclosed spot (2) represents the estrogen-stimulated 24K protein.

days of estradiol treatment. A major Coomassie-blue-stained spot is present at 24,000 daltons with a pI of 6.4. When this spot is excised and counted along with other prominent spots on the gel, a significant amount of radioactivity is detected with a ^{3}H/^{14}C ratio 2.2-fold above background (Table II). We estimate from these data that 24K in estrogen-stimulated cells represents about 1.6% of the total incorporation (^{3}H counts) into cytoplasmic protein, making 24K a major intracellular protein regulated by estrogen. In addition, 0.7% of newly synthesized protein occurs at 24K in unstimulated cells (^{14}C counts), indicating that the protein is synthesized constitutively.

In a further attempt to identify the 24K protein, we have compared the mobility of 24K on SDS-gels with that of the human milk proteins, casein and α-lactalbumin, and with the principal proteins found in cyst fluid obtained from patients with gross cystic breast disease (Haagensen et al., 1979). Figure 5 indicates that 24K does not comigrate on the gel with any of these proteins; furthermore, the size, cellular location, and amount of 24K suggest that it is not

Table II. Incorporation and ^{3}H/^{14}C Ratio for Individual
Proteins Resolved by Two-Dimensional Electrophoresis

Gel spot[a]	^{14}C dpm (control)	^{3}H dpm (estradiol)	^{3}H/^{14}C ratio	Percentage of baseline ratio[b]
1	54	897	16.57	250
2 (24K)	538	7900	<u>14.67</u>	222
3	237	2237	9.41	142
4	65	274	4.21	64
5	153	1027	6.72	101
6	628	3907	6.22	94
7	43	267	6.26	95
8	67	385	5.76	87
9	219	1313	6.00	91
10	235	1428	6.06	92
11	349	2831	8.11	122
12	81	633	7.85	118
13	72	534	7.38	111
14	658	4028	6.12	92
15	145	914	6.28	95
16	286	2052	7.17	108
17	672	5185	7.72	116
18	116	595	5.13	77
19	256	1604	6.27	95
20	186	919	4.94	75
21	111	995	8.99	135
Blank	25	153	<u>6.19</u>	94
			Avg. 6.64[c] ± 0.29 (S.E.)	

[a] The numbered spots shown in Fig. 6 were cut out and counted for ^{3}H- and ^{14}C-radioactivity.
[b] Baseline ratio is the average ratio of all numbered spots, excluding spots 1 and 2.
[c] Average of ratios excluding spots 1 and 2.

related to any of the other estrogen-regulated proteins reported for MCF-7 cells. It should be noted, however, that the size of 24K on SDS gels may be misleading. Under nondenaturing conditions, 24K appears in the excluded volume on Sephacryl S-300 columns indicating a much greater molecular weight ($\geqslant 10^6$ daltons; D. Adams, unpublished observation). This result could reflect a high degree of aggregation of the 24K molecule with itself or with another high-molecular-weight protein. The 24K protein could also be a degradation product produced during cell fractionation, although our preliminary experiments using various homogenization buffers and protease inhibitors, together with the cell-free translation data described below, argue against this possibility.

Like the identity of 24K, little is known about the mechanism of estrogenic regulation of this protein. Initial experiments suggest that transcriptional control

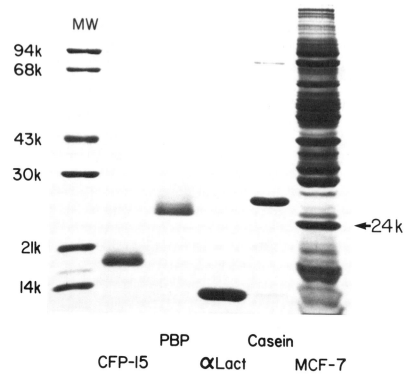

Figure 5. Coomassie blue staining pattern on SDS-polyacrylamide gels. From the left; molecular weight standards (MW); gross cystic disease fluid protein (GCDFP-15); progesterone binding protein (PBP) from gross breast cystic disease; human α-lactalbumin (αLact); human casein; MCF-7 double-label cytosol from 6 days estradiol stimulation.

may be involved. Using a double-label translation assay, we have shown that messenger RNA derived from estrogen-rescued MCF-7 cells encodes a major translation product of 24,000 mol. wt. (Fig. 6). This radioactive peak coincides with a prominent ratio peak, suggesting that estrogen-rescued cells have an increased rate of 24K synthesis. Whether this effect is due to an increase in the specific mRNA for 24K is not yet known. However, double-label analysis of 24K synthesis *in vivo* (Fig. 4) and *in vitro* (Fig. 6) both imply that a significant amount of 24K is produced in antiestrogen-treated control cells. Furthermore, the time course of the 24K response to estrogen stimulation measured in double-labeled cytosols parallels the pattern observed in cell-free translation assays (Fig. 7). At day 1, the $^3H/^{14}C$ ratio at 24K is only slightly increased above the baseline ratio. A progressive increase in the ratio then occurs, reaching maximum stimulation between 4 and 6 days of incubation with estradiol. The 24K response therefore appears to be a late effect of estrogen, similar to the time course of PgR stimulation and not the early response characteristic of IP(s). The cell-free translation data also indicate that the 24K ratio peak observed in MCF-7 cytosols is not due to protein modification or compartmentalization. Finally, these data suggest that estrogenic regulation of 24K probably does not represent induction of a dormant gene, but rather an increase in the relative abundance of 24K mRNA.

As mentioned previously, direct quantitation of the estrogen-regulated proteins in human breast cancer would require development of highly specific immunological assays. Our most recent work has been directed to this end and we have succeeded in preparing monoclonal antibodies to 24K using the hybridoma technique of

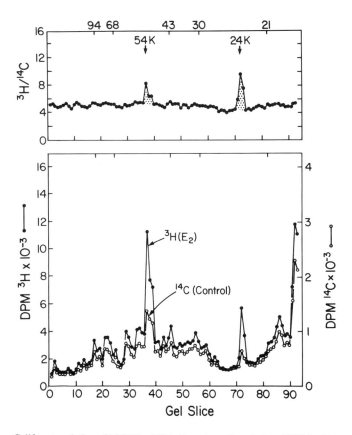

Figure 6.   Cellfree translation of MCF-7 mRNA. Two days after plating, MCF-7 cells were treated with nafoxidine (1 μM) for 6 days and then either continued on nafoxidine or treated with estradiol (10 nM) for 6 days. Messenger RNA was isolated from equal cell numbers in each treatment group, and equivalent amounts of mRNA were translated in the reticulocyte lysate system using [^{14}C]leucine (nafoxidine) or [^{3}H]leucine (estradiol) as tracer. Cellfree translation products were mixed and coelectrophoresed on SDS gels. The upper profile shows the ratio of ^{3}H/^{14}C in each gel slide; the lower profile indicates the total dpm for ^{3}H (●—●) and ^{14}C (○—○).

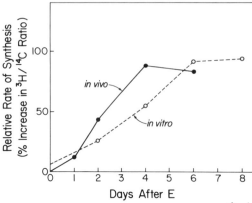

Figure 7. Time course of stimulation of the 24K protein. Cells were pretreated with nafoxidine and $^3H/^{14}C$ cytosols were prepared and analyzed as described in Fig. 1, except at the times of estrogen stimulation indicated. The degree of stimulation of the 24K protein was estimated by calculating the percentage increase in the $^3H/^{14}C$ ratio (●——●) at 24K over the baseline ratio. For comparative purposes, the time course for estradiol stimulation of the mRNA fraction (○-○) synthesizing 24K in a double-label cell-free translation assay is also included (taken from Adams et al., 1980).

Kohler and Milstein (1975). Preliminary experiments support our double-label ratio results, indicating that 24K is present in untreated or antiestrogen-treated MCF-7 cells and that estrogen treatment produces about a twofold increase in incorporation of [^3H]leucine into immunoprecipitable 24K. These antibodies are also able to detect 24K in the cytoplasm of MCF-7 cells growing in nude mice, using immunohistochemical staining techniques. The antibodies do not react with surrounding mouse tissues and can also specifically stain human breast carcinomas in paraffin-imbedded biopsy specimens. These monoclonal antibodies to 24K will therefore greatly facilitate the study of estrogen-regulated protein synthesis in human breast tumor cell lines and may improve the detection of hormone-responsive human breast tumors.

3.2.3. The 54K Nuclear Protein

A surprising result of the double-label translation assay shown in Fig. 6 was the appearance of another putative estrogen-regulated protein at 54,000 daltons. The seemingly large amount of messenger RNA for 54K observed in this experiment was puzzling because our in vivo results did not reveal a major ratio change at this molecular weight. One explanation for this anomaly is that 54K is not a cytoplasmic protein. Accordingly, our recent data indicate that 54K is the major newly synthesized protein present in MCF-7 nuclei (D. Adams, manuscript in preparation).

Observation of an estrogen-regulated 54K nuclear protein in human breast tumor cells is significant for several reasons. First, 54K may be related to a nucleolar 54K protein isolated by Chan et al. (1980) and detected in a broad range of human malignancies, including 94% of malignant breast tumors (Busch *et al.*, 1979). The protein is not found in benign tumors or normal tissue. In another model system, Crawford *et al.* (1981) have prepared a monoclonal antibody against a 53K phosphoprotein (p53) associated with SV40 viral T antigen. Again, this antibody detects a 53K protein in all cell lines derived from spontaneous tumors or from normal cells transformed by SV40 but not in normal cells, including mammary epithelial cells cultured from human milk. Similar results have been reported by Dippold *et al.* (1981) who also showed that p53 has kinase activity capable of phosphorylating serine and threonine but not tyrosine (Jay *et al.*, 1981). Interestingly, a number of viral "oncogenes" appear to code for 53–60K protein kinases that specifically phosphorylate tyrosine residues.

Although expression of a 53–54K protein now appears to be a common feature of malignant cells, it is probably characteristic of rapidly proliferating cells in general. For example, treatment of nondividing T lymphocytes with the mitogen, concanavalin A, also induces expression of a 53K protein (Milner and McCormick, 1980). Furthermore, normal kidney epithelium and fetal brain cells express high levels of 53K phosphoprotein during exponential growth but promptly shut down p53 synthesis after reaching contact inhibition of cell division (Dippold *et al.*, 1981). This result has obvious implications for our observation of a nuclear 54K ratio peak in MCF-7 cells undergoing growth rescue compared to antiestrogen-inhibited cells. Dippold *et al.* (1981) also find p53 antigenicity in the BT-20 (ER− PgR−) breast tumor cell line in addition to MCF-7, indicating that expression of this protein is not receptor-dependent. Our nuclear 54K ratio peak could then conceivably be due to estrogen-regulated protein translocation rather than *de novo* protein synthesis or could be a nonspecific effect of growth rescue. We are currently investigating these possibilities. Preliminary experiments indicate that 54K (and 24K) are in fact absent in the ER− PgR− cell line MDA-231. Although the exact relation of our nuclear 54K protein to similar proteins described by others remains unknown, the possibilities for 54K function in estrogen-regulated growth control promise to be an exciting avenue for future research.

4. Conclusions

Although we have made significant advances in our understanding of estrogen-sensitive breast tumors, we are far from unraveling hormone effects at the molecular level. Perhaps we have been too anxious to fit estrogen regulation of breast tumors into the mold established by classic effects of this hormone on target cells. Diversity is a hallmark of natural processes and it is quite possible that new modes of estrogen action will surface in breast tumor cells. We are now realizing that estrogen may not have a straightforward mechanism that impinges directly on breast tumor cells. Estrogen may act in concert with other hormones affecting a variety of cell types *in vivo* that respond in a coordinated fashion.

How then, shall we interpret estrogen regulation of specific protein synthesis in breast tumor cells under *in vitro* conditions? If we accept that current cell culture conditions do not reproduce the *in vivo* tumor environment (as is suggested by lack of estrogenic effects on tumor growth in culture), it is perhaps surprising that any estrogen-regulated protein observed in cultured tumor cells has relevance *in vivo*. Clearly, however, estrogen-stimulated protease and progesterone receptor activities in MCF-7 cells have correlates *in vivo*. For example, tamoxifen treatment can increase PgR in ER-positive cutaneous metastatic nodules taken from postmenopausal patients with breast cancer. This estrogenic property of a clinically important antiestrogen may improve progestagen effectiveness in counteracting the growth-promoting effects of estradiol in sensitive tumors (Namer *et al.,* 1980). As yet, no similar correlation can be made for the other hormone-regulated proteins described in breast tumor cell lines. Since biological activity-based assays are not available, development of highly specific antibodies, such as those recently prepared against 24K, will be required to detect the 46, 24, and 54K proteins *in vivo*. The advent of these immunological assays could very well bring a new level of quality control and sensitivity to the detection of hormone-dependent breast tumors.

Study of estrogenic effects in cultured breast tumor cells has therefore proved most valuable in identifying estrogen-regulated tumor markers. However, the old question of how estrogen stimulates tumor growth still remains unanswered. One might then ask whether estrogenic effects on tumor growth are necessarily related to hormone regulation of protein synthesis. A recent study has shown that estrogen regulates PgR levels in the MTW-9B rat mammary tumor but has no influence on tumor growth (Ip *et al.,* 1979). Similar observations were made in our laboratory using the DMBA-induced rat tumor model where a small number of tumors were autonomous in their growth but dependent upon estradiol for maintenance of PgR (Horwitz and McGuire, 1977b).

To sort out this question and others concerning the diverse effects and pathways of estrogen action in breast tumors will almost certainly require improvements in our cell culture models. A significant advance has already been made by development of defined media for the MCF-7 and ZR-75-1 cell lines (Barnes and Sato, 1979; Barnes, 1980; Allegra and Lippman, 1980). Removal of serum from culture medium permits an assessment of hormone function and interaction unclouded by the many growth factors known (and yet to be discovered) in serum. Furthermore, serum generally appears to stimulate cell proliferation and suppress differentiated functions which could lead to selection of cell populations that may not reflect those *in vivo* (Barnes and Sato, 1980; Orly *et al.,* 1980). Use of serumfree media will also permit detailed study of substratum and attachment factor requirements. There is increasing evidence that cellular morphology is determined by the extracellular matrix and that cellular shape is responsible for control of growth and function, including hormone responsiveness of mammary tumor cells (Gospodarowicz *et al.*, 1979; Yates and King, 1981). Advances in our understanding of cellular matrices could lead to successful culturing of normal mammary epithelial cells. Absence of hormone-responsive normal breast cell lines as control cultures is a notable deficiency in this field. Finally, a basic understanding of hormone-dependent tumor growth may require even more complex model systems. We may need to exploit tumor growth in athymic nude mice or coculture breast tumor cells with other cell lines. Indeed, at least one group has demonstrated a requirement for inoculation of pituitary tumor cells in addition to estradiol treatment to produce growth of MCF-7 cells consistently in nude mice (Leung and Shiu, 1981). Ideally, a system will be developed where estrogen-dependent tumor growth can be correlated with specific changes in the control of a hormone-dependent gene. A shift from inducible to constitutive expression of a gene product involved in growth control is one possible mechanism that may apply to breast cancer. Clearly, the future holds many interesting possibilities for defining the role of estrogen and other steroid hormones in regulating both normal and neoplastic human mammary tissue.

REFERENCES

Adams, D. J., Edwards, D. P., and McGuire, W. L., 1980, Estrogen regulation of specific messenger RNA's in human breast cancer cells, *Biochem. Biophys. Res. Commun.* **97**:1354.

Allegra, J. C., and Lippman, M. E., 1978, Growth of a human breast cancer cell line in serum-free hormone-supplemented medium, *Cancer Res.* **38**:3823.

Baker, H. J., and Shapiro, D. J., 1977, Kinetics of estrogen induction of *Xenopus laevis* vitellogenin messenger RNA as measured by hybridization to complementary DNA, *J. Biol. Chem.* **252**:8428.

Barker, K. L., Adams, D. J., and Donohue, T. M., 1981, Regulation of the levels of mRNA for glucose-6-phosphate dehydrogenase and its rate of translation in the uterus by estradiol, in: *Cellular and Molecular Mechanisms of Implantation* (S. Glasser and D. Bullock, eds.), pp. 269–281, Plenum Press, New York.

Barnes, D., 1980, Factors that stimulate proliferation of breast cancer cells *in vitro* in serum-free medium, in: *Cell Biology of Breast Cancer* (C. McGrath, M. Brennan, and M. Rich, eds.), pp. 227–287, Academic Press, New York.

Barnes, D., and Sato, G., 1979, Growth of a human mammary tumor cell line in a serum-free medium, *Nature* **281**:388.

Barnes, D., and Sato, G., 1980, Serum-free cell culture: A unifying approach, *Cell* **22**:649.

Baulieu, E. E., Alberga, A., Raynaud-Jammet, C., and Wira, C. R., 1972, New look at the very early steps of oestrogen action in uterus, *Nature New Biol.* **236**:236.

Beatson, G. T., 1896, On the treatment of inoperable cases of carcinoma of the mamma: Suggestions for a new method of treatment, with illustrative cases, *Lancet* **2**:104.

Bolander, F. F., Jr., and Topper, Y. J., 1979, Stimulation of lactose synthetase activity and casein synthesis in mouse mammary explants by estradiol, *Endocrinology* **106**:490.

Bronzert, D. A., Monaco, M. E., Pinkus, L., Aitken, S., and Lippman, M. E., 1981, Purification and properties of estrogen-responsive cytoplasmic thymidine kinase from human breast cancer, *Cancer Res.* **41**:604.

Burke, R. E., Harris, S. C., and McGuire, W. L., 1978, Lactate dehydrogenase in estrogen-responsive human breast cancer cells, *Cancer Res.* **38**:2773.

Busch, H., Gyorkey, F., Busch, R. K., Davis, F. M., Gyorkey, R., and Smetna, A., 1979, A nucleolar antigen found in a broad range of human malignant tumor specimens, *Cancer Res.* **39**:3024.

Butler, W. B., Kirkland, W. L., and Jorgensen, T. L., 1979, Induction of plasminogen activator by estrogen in a human breast cancer cell line (MCF-7), *Biochem. Biophys. Res. Commun.* **90**:1328.

Chambon, P., Benoist, C., Breathnach, R., Cochet, M., Gannon, F., Gerlinger, P., Krust, A., Lemeur, M., LePennec, J. P., Mandel, J. L., O'Hare, K. U., and Perrin, F., 1979, Structural organization and expression of ovalbumin and related chicken genes, in: *From Gene to Protein: Information Transfer in Normal and Abnormal Cells* (T. R. Russell, K. Brew, H. Faber, and J. Schultz, eds.), pp. 55–83, Academic Press, New York.

Chan, L., and O'Malley, B. W., 1976, Mechanism of action of the sex steroid hormones, *N. Engl. J. Med.* **294**:1322.

Chan, P.-K., Feyerabend, A., Busch, R. K., and Busch, H., 1980, Identification and partial purification of human tumor nucleolar antigen 54/6.3, *Cancer Res.* **40**:3194.

Crawford, L. V., Pim, D. C., Gurney, E. G., Goodfellow, P., and Taylor-Papadimitriou, J., 1981, Detection of a common feature in several human tumor cell lines—a 53,000 dalton protein, *Proc. Natl. Acad. Sci. U.S.A.* **78**:41.

Cunningham, D. D., Carney, D. H., and Glenn, K. C., 1979, A cell-surface component involved in thrombin-stimulated cell division, in: *Hormones and Cell Culture, Book A* (G. H. Sato and R. Ross, eds.), pp. 199–218, Cold Spring Harbor Laboratory, Cold Spring Harbor, New York.

Deeley, R. G., Gordon, J. I., Burns, A. T. H., Mullinix, K. P., Bina-Stein, M., and Goldberger, R. F., 1977, Primary activation of the vitellogenin gene in the rooster, *J. Biol. Chem.* **252**:8310.

Dippold, W. G., Jay, G., DeLeo, A. B., Khoury, G., and Old, L. J., 1981, p53 Transformation-related protein: Detection by monoclonal antibody in mouse and human cells, *Proc. Natl. Acad. Sci. U.S.A.* **78**:1695.

Edwards, D. P., Chamness, G. C., and McGuire, W. L., 1979, Estrogen and progesterone receptor proteins in breast cancer, *Biochim. Biophys. Acta* **560**:457.

Edwards, D. P., Adams, D. J., and McGuire, W. L., 1980a, Estrogen induced synthesis of specific proteins in human breast cancer cells, *Biochem. Biophys. Res. Commun.* **93**:804.

Edwards, D. P., Murthy, S. R., and McGuire, W. L., 1980b, Effects of estrogen and antiestrogen on DNA polymerase in human breast cancer, *Cancer Res.* **40**:1722.

Edwards, D. P., Adams, D. J., and McGuire, W. L., 1981, Estradiol stimulated synthesis of a major intracellular protein in human breast cancer cells (MCF-7), *Breast Cancer Treat. Res.* **1**:209.

Folca, P. J., Glascock, R. F., and Irvine, W. T., 1961, Studies with tritium-labeled hexestrol in advanced breast cancer, *Lancet* **2**:796.

Goldman, R. D., Kaplan, N. O., and Hall, T. C., 1964, Lactate dehydrogenase in human neoplastic tissues, *Cancer Res.* **24**:389.

Gospodarowicz, D., Vlodavsky, I., Greenburg, G., and Johnson, L. K., 1979, Cellular shape is determined by the extracellular matrix and is responsible for the control of cellular growth and function, in: *Hormones and Cell Culture, Book B* (G. H. Sato and R. Ross, eds.), pp. 561–592, Cold Spring Harbor Laboratory, Cold Spring Harbor, New York.

Haagensen, D. E., Mazoujian, G., Dilley, W. G., Pedersen, C. E., Kister, S. J., and Wells, S. A., 1979, Breast gross cystic disease fluid analysis. I. Isolation and radioimmunoassay for a major component protein, *J. Natl. Cancer Inst.* **62**:239.

Hakim, A. A., 1980, Estradiol-induced biochemical changes in human neoplastic cells: Estradiol-mediated protease, *Cancer Biochem. Biophys.* **4**:173.

Hilf, R., Rector, W. D., and Orlando, R. A., 1976, Multiple molecular forms of lactate dehydrogenase and glucose 6-phosphate dehydrogenase in hormonal and abnormal human breast tissues, *Cancer* **37**:1825.

Hornebeck, W., Brechemier, D., Bellon, G., Adnet, J. J., and Robert, L., 1980, Biological significance of elastase-like enzymes in arteriosclerosis and human breast cancer, in: *Proteinases and Tumor Invasion* (A. J. Barrett, A. Baici, and P. Strauli, eds.), pp. 117–141, Raven Press, New York.

Horwitz, K. B., and McGuire, W. L., 1977a, Estrogen control of progesterone receptor in human breast cancer, *J. Biol. Chem.* **253**:2223.

Horwitz, K. B., and McGuire, W. L., 1977b, Progesterone and progesterone receptors in experimental breast cancer, *Cancer Res.* **37**:1733.

Horwitz, K. B., and McGuire, W. L., 1978, Nuclear mechanisms of estrogen action: Effects of estradiol and antiestrogens on estrogen receptors and nuclear receptor processing, *J. Biol. Chem.* **253**:8185.

Horwitz, K. B., McGuire, W. L., Pearson, O. H., and Segaloff, A., 1975, Predicting response to endocrine therapy in human breast cancer, *Science* **189**:726.

Horwitz, K. B., Zava, D. T., Thilagar, A. K., Jensen, E. M., and McGuire, W. L., 1978, Steroid receptor analyses of nine human breast cancer cell lines, *Cancer Res.* **38**:2434.

Horwitz, K. B., Aiginger, P., Kuttenn, F., and McGuire, W. L., 1981, Nuclear estrogen receptor release from antiestrogen suppression: Amplified induction of progesterone receptor in MCF-7 human breast cancer cells, *Endocrinology* **108**:1703.

Huggins, C., and Bergenstal, D. M., 1951, Inhibition of human mammary and prostatic cancers by adrenalectomy, *Cancer Res.* **12**:134.

Huggins, C., and Dao, T. L.-Y., 1954, Characteristics of adrenal-dependent mammary cancers, *Ann. Surg.* **140**:497.

Ip, M., Milholland, R. J., and Rosen, F., 1979, Mammary cancer: Selective action of the estrogen receptor complex, *Science* **203**:361.

Ivarie, R. D., and O'Farrell, P. H., 1978, The glucocorticoid domain: steroid-mediated changes in the rate of synthesis of rat hepatoma proteins, *Cell* **13**:41.

Jay, G., Khoury, G., DeLeo, A. B., Dippold, W. G., and Old, L. J., 1981, p53 Transformation-related protein: Detection of an associated phosphotransferase activity, *Proc. Natl. Acad. Sci. U.S.A.* **78**:2932.

Jensen, E. V., 1981, Hormone dependency of breast cancer, *Cancer* **47**:2319.

Jensen, E. V., DeSombre, E. R., and Jungblut, P. W., 1967, Estrogen receptors in hormone-responsive tissues and tumors, in: *Endogenous Factors Influencing Host–Tumor Balance* (R. W. Wissler, T. L. Dao, and S. Wood, Jr., eds.), pp. 15–30, University of Chicago Press, Chicago.

Kaye, A. M., and Reiss, N., 1980, The uterine "estrogen induced protein" (IP): Purification, distribution and possible function, in: *Steroid Induced Uterine Proteins* (M. Beato, ed.), pp. 3–19, Elsevier/North-Holland Biomedical Press, New York.

Kaye, A. M., Reiss, N., Iacobelli, S., Bartoccioni, E., and Marchetti, P., 1980, The "estrogen-induced protein" in normal and neoplastic cells, in: *Hormones and Cancer* (S. Iacobelli, H. R. Lindner, R. J. B. Kino, and M. E. Lippman, eds.), pp. 41–51, Raven Press, New York.

King, R. J. B., 1979, How important are steroids in regulating the growth of mammary tumors?, in: *Biochemical Actions of Hormones*, Vol. VI, pp. 247–264, Academic Press, New York.

Knight, W. A., Livingston, R. B., Gregory, E. J., and McGuire, W. L., 1977, Estrogen receptor as an independent prognostic factor for early recurrence in breast cancer, *Cancer Res.* **37**:4669.

Kohler, G., Milstein, C., 1975, Derivation of specific antibody-producing and tumor cell lines by cell fusion, *Eur. J. Immunol.* **6**:511.

Leung, C. K. H., and Shiu, R. P. C., 1981, Required presence of both estrogen and pituitary factors for the growth of human breast cancer cells in athymic nude mice, *Cancer Res.* **41**:546.

Lockwood, D. H., Boytovich, A. E., Stockdale, F. E., and Topper, Y. J., 1967, Insulin-dependent DNA polymerase and DNA synthesis in mammary epithelial cells *in vitro*, *Proc. Natl. Acad. Sci. U.S.A.* **58**:658.

Mairesse, N., Devleeschouwer, N., Leclercq, G., and Galand, P., 1980, Estrogen-induced protein in the human breast cancer cell line MCF-7, *Biochem. Biophys. Res. Commun.* **97**:1251.

Mairesse, N., Devleeschouwer, N., Leclercq, G., and Galand, P., 1981, Estrogen-induced protein in the human breast cancer cell line MCF-7: Further characterization. Fifth International Symposium of the Journal of Steroid Biochemistry, Puerto Vallarta, Jalisco, Mexico.

Martin, T. J., Findlay, D. M., MacIntyre, I., Eisman, J. A., Michelangeli, V. P., Moseley, J. M., and Partridge, N. C., 1980, Calcitonin receptors in a cloned human breast cancer cell line (MCF-7), *Biochem. Biophys. Res. Commun.* **96**:150.

McGuire, W. L., Carbone, P. P., Sears, M. E., and Escher, G. C., 1975, Estrogen receptors in human breast cancer: An overview, in: *Estrogen Receptors in Human Breast Cancer* (W. L. McGuire, P. P. Carbone, and E. P. Vollmer, eds.), pp. 1–7, Raven Press, New York.

McKnight, S. G., and Palmiter, R. D., 1979, Transcriptional regulation of the ovalbumin and conalbumin genes by steroid hormones in the chick oviduct. *J. Biol. Chem.* **254**:9050.

Miller, L. K., Tuazon, F. B., Niu, E.-M., and Sherman, M. R., 1981, Human breast tumor estrogen receptor: Effects of molybdate and electrophoretic analyses, *Endocrinology* **108**:1369.

Milner, J., and McCormick, F., 1980, Lymphocyte stimulation: Concanavalin A induces the expression of a 53K protein, *Cell Biol. Int. Rep.* **4**:663.

Namer, M., Lalanne, C., and Baulieu, E. E., 1980, Increase of progesterone receptor by tamoxifen as a hormonal challenge test in breast cancer, *Cancer Res.* **40**:1750.

Notides, A., and Gorski, J., 1966, Estrogen-induced synthesis of a specific uterine protein, *Proc. Natl. Acad. Sci. U.S.A.* **56**:230.

O'Farrell, P. J., 1975, High resolution two-dimensional electrophoresis of proteins, *J. Biol. Chem.* **250**:4007.

Orly, J., Sato, G., and Erickson, G. F., 1980, Serum suppresses the expression of hormonally induced functions in cultured granulosa cells, *Cell* **20**:817.

Osborne, C. K., Yockmowitz, M. G., Knight, W. A., and McGuire, W. L., 1980, The value of estrogen and progesterone receptors in the treatment of breast cancer, *Cancer* **46**:2884.

Paranjpe, M., Engel, L., Young, N., and Liotta, L. A., 1980, Activation of human breast carcinoma collagenase through plasminogen activator, *Life Sci.* **26**:1223.

Poole, A. R., Recklies, A. D., and Mort, J. S., 1980, Secretion of proteinases from human breast tumors: Excessive release from carcinomas of a thiol proteinase, in: *Proteinases and Tumor Invasion* (A. J. Barrett, A. Baici and P. Strauli, eds.), pp. 81–95, Raven Press, New York.

Quigley, J. P., Goldfarb, R. H., Scheiner, C., O'Donnell-Tormey, J., and Yeo, T. K., 1980, Plasminogen activator and the membrane of transformed cells, in: *Tumor Cell Surfaces and Malignancy* (R. O. Hynes and C. F. Fox, eds.), pp. 773–796, Alan R. Liss, Inc., New York.

Reiss, N., and Kaye, A. M., 1981, Identification of the major component of the estrogen-induced protein of rat uterus as the BB isoenzyme of creatine-kinase, *J. Biol. Chem.* **256**:5741.

Shafie, S. M., 1980, Estrogen and the growth of breast cancer: New evidence suggests indirect action, *Science* **209**:701.

Shafie, S. M., and Liotta, L. A., 1980, Formation of metastasis by human breast carcinoma cells (MCF-7) in nude mice, *Cancer Lett.* **11**:81.

Sherman, M. R., Tuazon, F. B., and Miller, L. K., 1980, Estrogen receptor cleavage and plasminogen activation by enzymes in human breast tumor cytosol, *Endocrinology* **106**:1715.

Singhakowinta, A., Potter, H. G., Buroker, T. R., Samel, B., Brooks, S. C., and Vaitkervicius, V. K., 1976, Estrogen receptor and natural course of breast cancer, *Ann. Surg.* **183**:84.

Sirbasku, D. A., 1980, Estromedins: Uterine-derived growth factors for estrogen-responsive tumor cells, in: *Control Mechanisms in Animal Cells* (A. Shields, R. Levi-Montalcini, S. Iacobelli, and L. Jimenez de Asua, eds.), pp. 293–298, Raven Press, New York.

Sirbasku, D. A., and Benson, R. H., 1980, Proposal of an indirect (estromedin) mechanism of estrogen-induced mammary tumor cell growth, in: *Cell Biology of Breast Cancer* (C. McGrath, M. Brennan, and M. Rich, eds.), pp. 289–314, Academic Press, New York.

Skipper, J. K., Eakle, S. D., and Hamilton, T. H., 1980, Modulation by estrogen of synthesis of specific uterine proteins, *Cell* **22**:69.

Stewart, J. F., King, R. J. B., Sexton, S. A., Millis, R. R., Rubens, R. D., and Haward, J. L., 1981, Oestrogen receptors, sites of metastatic disease and survival in recurrent breast cancer, *Eur. J. Cancer* **17**:449.

Supowit, S. C., and Rosen, J. M., 1980, Gene expression in normal and neoplastic mammary tissue, *Biochem.* **19**:3432.

Tata, J. R., 1979, Control by oestrogen of reversible gene expression: The vitellogenin model, *J. Steroid Biochem.* **11**:361.

Terenius, L., 1968, Selective retention of estrogen isomers in estrogen-dependent breast tumors of rats demonstrated by *in vitro* methods, *Cancer Res.* **28**:328.

Toft, D., and Gorski, J., 1966, A receptor molecule for estrogens: Isolation from the rat uterus and preliminary characterization, *Proc. Natl. Acad. Sci. U.S.A.* **55**:1574.

Topper, Y. J., and Freeman, C. S., 1980, Multiple hormone interactions in the developmental biology of the mammary gland, *Physiol. Rev.* **60**:1049.

Vignon, F. V., Terqui, M., Westley, B., Ducoq, D., and Rochefort, H., 1980, Effects of plasma estrogen sulfate in mammary cancer cells, *Endocrinology* **106**:1079.

Westley, B., and Rochefort, H., 1979, Estradiol induced proteins in the MCF-7 human breast cancer cell line, *Biochem. Biophys. Res. Commun.* **90**:410.

Westley, B., and Rochefort, H., 1980, A secreted glycoprotein induced by estrogen in human breast cancer cell lines, *Cell* **20**:353.

Yates, J., and King, R. J. B., 1981, Correlation of growth properties and morphology with hormone responsiveness of mammary tumor cells in culture, *Cancer Res.* **41**:258.

Zava, D. T., and McGuire, W. L., 1978, Androgen action through the estrogen receptor in a human breast cancer cell line, *Endocrinology* **103**:624.

2

Modulation of an Estrogen-Induced Protein in the MCF₇ Human Mammary Cancer Cell

Henri Rochefort

1. Introduction

Estrogens are able to induce two series of responses in a given target tissue. First, they increase cell proliferation and stimulate the general metabolism of the cells by favoring their entry into an active G_1 step of the cell cycle (pleiotypic effect) (Herschko *et al.*, 1971). Secondly, they stimulate the biosynthesis of specific proteins. The mechanism of the pleiotypic effect is poorly understood and may involve common intermediary steps between the binding of the receptor in the nucleus and the final responses. In any case, this pleiotypic effect can also be triggered by other hormones and is therefore not specific for estrogens.

The modulation of cell proliferation by estrogens and antiestrogens is certainly the most important response to understand and to control in the treatment of hormone-dependent cancer, for instance. However, we will not consider this problem here but will concentrate on the basic problem of the regulation by estrogens of the expression of a limited number of genes in breast cancer cells, giving rise to the induction of specific proteins. Regulation of gene expression by estrogen has been mostly studied in birds where induced proteins are highly abundant. By contrast, in mammals, we lack good marker proteins to study the

Henri Rochefort ● Unité d'Endocrinologie Cellulaire et Moléculaire, U 148 INSERM, 34100 Montpellier, France.

mechanism of action of estrogens and antiestrogens. The major mammalian estrogen-induced proteins which have been described, such as the progesterone receptor (Horwitz and McGuire, 1978) and the uterine-induced protein (Notides and Gorski, 1966), are intracellular.

We will describe here other estrogen-induced proteins which are released by breast cancer cells in culture (Westley and Rochefort, 1980). With this system, we will discuss how one can specify the nature of the receptor that mediates the effect of one given steroid hormone and whether the specificity of the response is located at the hormone or at the receptor level (Rochefort *et al.*, 1981b). We will then discuss, through the study of estrogen antagonists, the importance of a normal activation of the estrogen receptor (ER) and of the integrity of the cellular DNA in obtaining a normal hormone-regulated expression of these proteins.

These studies have been mostly performed in the human breast cancer cell line MCF_7, which has been studied in several laboratories for eight years and which is estrogen-sensitive (Lippman *et al.*, 1976; McGuire *et al.*, 1980). Although this is debated (Sirbasku and Benson, 1979; Sonnenschein and Soto, 1980), estrogens stimulate *in vitro* the proliferation of breast cancer cells (Lippman *et al.*, 1976; Chalbos *et al.*, 1981). Antiestrogens block the proliferation of MCF_7 cells (McGuire *et al.*, 1980) when the ER is not occupied by estrogens, but are able partially to stimulate the synthesis of progesterone receptor (PR) (Horwitz and McGuire, 1978).

2. Estrogen-Induced Proteins Released by MCF₇ Cells in Culture

In order to improve our understanding of the mechanism by which estrogens and antiestrogens regulate gene expression in breast cancer cells, we have first looked at other estrogen-induced proteins in human breast cancer cell lines. The cell culture approach facilitates labeling of proteins and nucleic acid, in cells living in conditions close to the *in vivo* situation.

The MCF_7 cells were first withdrawn from endogenous estrogens in a medium containing 10% charcoal-treated serum; they were then cultured in the same medium containing different hormones or antihormones and were finally labeled with [^{35}S]methionine. The labeled proteins of the cell lysate and of the medium were then analyzed by sodium dodecyl sulfate (SDS)-polyacrylamide gel electrophoresis. The fluorograms of these gels revealed different bands, the intensity of which could be quantified in a gel scanner. We found no clear effect of estradiol on the labeling of intracellular proteins by this technique. Using the two-dimensional technique of O'Farrell (1975), we found three spots consistently stimulated by estradiol. However, the most dramatic effect of estradiol was seen for the secreted proteins: after 12 hr of treatment, a protein of molecular weight 52,000 (52K)* was induced by estradiol (E_2) (Westley and Rochefort, 1979).

Similar E_2-induced proteins were found released into the medium by two other human breast cancer cell lines, ZR_{75-1} and $T_{47}D$. By contrast, two cell lines containing no ER (BT_{20} and HL_{100}) showed no effect of E_2 on secretory proteins (Westley and Rochefort, 1980). Several pieces of evidence indicate that the 52K protein in MCF₇ cells is a glycoprotein: microheterogeneity in two-dimensional SDS-PAGE, pH between 5.5 and 6.5, sensitivity to neuraminidase and tunicamycin, [³H]fucose labeling, and binding to Con A Sepharose (F. Veith and M. Garcia, unpublished results). The glycoprotein made up 30 to 40% of the total secreted proteins. The purification of the 52K protein, with the aim of producing specific antibodies, is in progress.

The mechanism by which E_2 stimulates its biosynthesis is presently unknown. However, we have indirect evidence that accumulation of mRNA is involved since actinomycin D and α-amanitin prevent its induction without modifying the noninduced proteins (Westley *et al.*, 1980). It is unlikely that DNA synthesis and cell differentiation are required for the induction of the 52K protein, since Ara C, which inhibits DNA synthesis, did not prevent the stimulation of the 52K protein by E_2. Adams *et al.* (1980) have shown, using *in vitro* translation of poly A + RNA from MCF₇ cells treated or untreated by estrogens, that E_2 (after blockade by antiestrogen) was able to increase the concentration of mRNA for a 54K and a 23K protein. We then studied the hormone specificity of induction of the 52K protein (Westley and Rochefort, 1980). When increasing concentrations of E_2 were used, a progressive increase of 52K synthesis was observed. The half-maximal induction (ED_{50}) was $\simeq$ 10 pM E_2 and near-optimal induction was obtained with 0.1 nM E_2. The discrepancy observed between the ED 50 of E_2 and its K_D for the ER might be real, thus indicating that the activation of a fraction of the ER is sufficient to provoke maximal effect. However, it might very well be apparent since the real affinity of E_2 for its receptor is most likely higher than that measured by Scatchard plot and the actual concentration of unbound E_2 in the cell at 37°C has not been measured. The dose–response curve for 52K induction was in any case consistent with a progressive occupation and activation of the ER by E_2. The induction was found to be specific for ER ligands since estrone and estriol were 10 times less efficient than E_2 while progesterone, dexamethasone, prolactin, and T_3 were totally inactive when tested in the absence of estrogen.

The androgen dihydrotestosterone (DHT) was also found to be inactive at

* The mol. wt. was first found to be 46,000 daltons by using 15% gel polyacrylamide (Westley and Rochefort, 1979). It seems to be closer to 52,000 daltons in 10% gel polyacrylamide and in the marker proteins provided by New England Nuclear.

physiological concentrations that occupy and activate the androgen receptor. However, at pharmacological concentrations, DHT, which is not transformed into estrogen, was found to increase the biosynthesis of the 52K estrogen-induced protein as already shown in the rat uterus for induced proteins (IP) (Ruh and Ruh, 1975; Garcia and Rochefort, 1977). Also some adrenal androgens that have a much higher affinity for the ER, such as 5-androstene-3β,17β-diol and 5α-androstane-3β,17β-diol, were able to induce this protein at close to physiological concentrations (Adams et al., 1981). For all ER ligands, we found that their relative ability to induce the 52K protein was proportional to their affinity for the ER.

We concluded that the 52K secretory protein was induced specifically by the ER–estrogen complex. However, the specificity was given by the receptor protein molecule rather than by the steroid molecule, since other steroids such as androgens could fully activate the ER to give the same specific response. The informational molecule at the chromatin level appears therefore to be the protein rather than the hormone, whose function is mostly to trigger and maintain an activated state of the molecule. These results were also found for other estrogen target tissues and other induced proteins (Garcia and Rochefort, 1978; Zava and McGuire, 1978). They strongly suggest specific interaction(s) of the receptor complexes with genomic material. The case of the nonsteroidal antiestrogen is an interesting exception since these ER ligands are able to activate and translocate the ER into the nucleus (Clark et al., 1976; Rochefort et al., 1980) giving a nuclear ER–antiestrogen complex which is, however, unable fully to provoke estrogen-specific responses.

3. Dissociated Effect of Nonsteroidal Antiestrogens

We believe that to study the mechanisms by which one can specifically inhibit the efficiency of the estrogen receptor, whether or not it is exposed to estrogen, is interesting in improving our understanding of the molecular mechanism of hormone action as well as that of hormone resistance. We will review here recent data from our laboratory dealing with two classes of antiestrogens, one (tamoxifen; Tam) works via its interaction with the ER and the other (bromodeoxyuridine; BrdU) modifies the structure of DNA and therefore acts at the postreceptor level. Nonsteroidal antiestrogens prevent tumor proliferation and they are extensively used in breast cancer treatment for this purpose. We have studied the specificity of action of several tamoxifen analogues and derivatives on the MCF$_7$ cell growth. We found a reasonably good correlation between the affinity of tamoxifen derivatives for the ER and their efficiency to prevent cell growth. For instance, the growth of MCF$_7$ cells is inhibited by tamoxifen and rescued by estrogen (Lippman *et al.*, 1976; Coezy *et al.*, 1982). We have recently shown that the 4-hydroxy tamoxifen (OHT) was 100 times more effective than Tam in preventing cell growth and that higher concentrations of $^3HE_2$ than tamoxifen were required to rescue the cells (Coezy *et al.*, 1982).

This was consistent with the 200-fold higher affinity of OHT for ER (Borgna and Rochefort, 1980) as compared to Tam. We will not discuss the mechanism of this effect here and will concentrate on the effect of antiestrogen on the synthesis of the 52K induced proteins described in MCF₇. In these cells OH Tam and Tam are both able to induce the PR (F. Vignon, unpublished data). In this respect, nonsteroidal antiestrogens behave as partial estrogen agonists. However, to our surprise, we found that in the same cells, both Tam and OH Tam were unable to stimulate the synthesis of any secretory proteins (Westley and Rochefort, 1980). Thus, when added with E_2, they fully prevented the induction of the 52K protein by either E_2 or high doses of DHT (Rochefort *et al.*, 1980). OH Tam was 10 times more potent than Tam in blocking the induction of the secreted protein by E_2. We have checked that Tam and OH Tam were not interconverted by the cells. The molar ratio Tam/E_2 necessary for 50% inhibition of the E_2-induction of the 52K protein varied from 10^4 to 10^2, whether Tam was added before or together with E_2 (F. Vignon, unpublished data). This was consistent with a slower rate of entry of Tam into the cells as compared to E_2 (Rochefort and Capony, 1973).

We can draw several conclusions from these studies (Fig. 1).

1. Tam is active by itself as a full antagonist for the 52K protein and as a partial agonist–antagonist for the PR.
2. OH Tam is at least 20 times more potent than Tam in preventing cell growth. Since we know that OH Tam is formed *in vivo* from Tam and accumulates at the nuclear ER level (at least in the rat uterus and chicken oviduct) (Borgna and Rochefort, 1981), we propose that this metabolite may also be active *in vivo* in human breast cancer cells.
3. The 52K protein appears to be more closely related to cell growth than the PR, since it is inhibited by antiestrogens which blocked cell growth but induced PR.
4. The effects of antiestrogens in these cells are dissociated since these drugs are partial agonists for the PR and full antagonist for the 52K glycoprotein. This suggests that the nuclear ER Tam (or OH Tam) complex is able to induce some gene products such as PR, but not all (e.g., the 52K protein). This dissociated effect of antiestrogens on the regulation of two different genes has also been shown in chicken liver where Tam is able partially to induce in liver apolipoprotein II and vitellogenin (Blue and Williams, 1981) but not apoprotein B (Capony and Williams, 1981) and ovalbumin (Sutherland *et al.*, 1977).

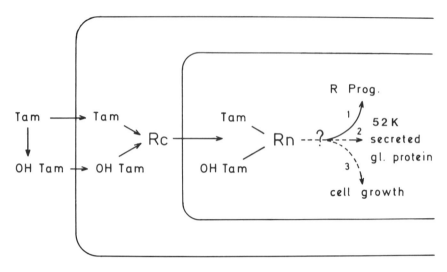

Figure 1.   Interaction and mechanism of action of Tam and OH Tam in MCF$_7$ cells via the cytosol (Rc) and nuclear (Rn) estrogen receptor. When tested alone, both antiestrogens tamoxifen (Tam) and hydroxytamoxifen (OH Tam) are active in MCF$_7$ cells; they induce the nuclear translocation of ER and provoke three types of responses (1,2,3). Tam is a partial agonist in pathway 1, a full antagonist in pathway 2, and might be a full antagonist on cell growth in pathway 3 since it has been shown that estrogens stimulate cell proliferation (Lippman et al., 1976; Chalbos et al., 1981).

The reason for this dissociated effect observed in MCF$_7$ cells is unknown since we do not know whether the altered production of the 52K protein in the medium after Tam treatment is due to a lack of induction of the 52K messenger RNA or to a lack of secretion of the protein (Palmiter et al., 1978).

4.   Altered Activation of ER by Antiestrogens

It is known that the synthetic antiestrogens both prevent specific estrogen responses such as induced protein and decrease breast cancer cell growth. The mechanism of the antitumor effect of these drugs may be more complex than a simple antiestrogenic effect; it may involve interactions with binding entities other than the ER (Sutherland et al., 1980) and will not be considered here. We will thus restrict this discussion to the mechanism by which antiestrogens prevent estrogen-specific responses.

To understand the mechanism of estrogen action, it is important to know why the binding to ER of an antiestrogen such as Tam or OH Tam is not as efficient as that of E$_2$ in terms of biological response. It had been proposed that antiestrogens have a decreased activity since they dissociate much more rapidly from the ER than estrogens (Bouton and Raynaud, 1978). This is true for the administered compounds such as tamoxifen, nafoxidine, Ci 628. However, *in vivo,* these compounds are transformed into higher-affinity metabolites bearing a phenol function which do not display any apparent difference of binding to ER as compared to E_2 (Borgna and Rochefort, 1980). For instance, OH Tam does not dissociate more rapidly from ER compared to E_2. We are then faced with the following dilemma: Either it should be possible to find a difference of ER activation by estrogen and antiestrogen, in which case an effect of antiestrogens mediated by the ER would be consistent, or it should not be possible to find such a difference, implying that the antiestrogenic effect of these drugs could be mediated by another binding entity which would inhibit the efficiency of the activated ER complex at a further step.

We have been able to find such a defective ER activation by antiestrogens which can be detected *in vitro* by comparing their dissociation rates from "native" ER stabilized by molybdate and from activated ER (Rochefort and Borgna, 1981). While the dissociation rate constant $(K-)$ of E_2 is markedly decreased after ER activation (Shyamala and Leonard, 1980), that of OH Tam is not modified. These results support the idea that antiestrogens are able to activate ER for nuclear translocation and DNA binding but not for full estrogenic activity. We therefore propose that the criterion for a full activation of ER is not the occurence of ER binding to DNA or to nuclei but rather an increased stability of the ER–ligand complex as evidenced by a decreased $K-$ following ER activation. This criterion may be useful for screening putative antiestrogens. It might also reflect a different conformational change of the hormone-binding site in the ER depending on whether it is activated by estrogens or antiestrogens. More recently, we have found two other criterion of receptor activation which could discriminate between the effects of estrogen and antiestrogen binding on the ER. These are the apparent lower affinity of ER for DNA after activation with OH Tam (E. Evans *et al.,* 1982) and a different interaction of a monoclonal antibody to the calf uterine ER (J. L. Borgna and Rochefort, unpublished). The molecular mechanism to explain the relative inefficiency of the ER–antiestrogen complex in chromatin to stimulate gene expression remains to be specified.

5. BrdU Incorporation into DNA Inhibits Estrogen-Induced Responses

There are several specific ways to inhibit the synthesis of estrogen-induced proteins (Fig. 2); the classic antiestrogens act as competitive inhibitors on the hormone-binding sites, molybdate can prevent ER activation (Shyamala and Leonard, 1980), and intercalating agents can inhibit ER binding to DNA and to the nucleus (André *et al.,* 1977). We have found another type of antiestrogen that might block the effect of E_2 at the acceptor chromatin sites and not at the ER site. When the MCF₇ cells are grown with 5 μg/ml of 5 bromodeoxyuridine

(BrdU) in the dark, the induction of the 52K protein by E_2 is totally inhibited (from 1 day to 4 days treatment) (Garcia *et al.*, 1981). This inhibition is specific for the E_2-induced proteins, since it is observed for the 52K and the PR but not for other proteins not regulated by E_2. This antiestrogenic effect of BrdU appears to depend on its incorporation into DNA since (1) it requires a lag of 24 hr, (2) ARA-C (a DNA synthesis inhibitor) inhibits this effect, and (3) thymidine, which competes with BrdU for DNA incorporation, also inhibits its effect. Conversely, deoxycytidine does not prevent this effect, suggesting that the mechanism of action of BrdU in this case is not due to a decrease in the free nucleotide pool. This effect of BrdU is an example of an inhibition of steroid hormone action which is probably located at the DNA level. It is also in agreement with the fact that the BrdU-substituted DNA has a higher *in vitro* affinity than native DNA for the ER (Kallos *et al.*, 1978). Of the many possibilities, one particularly attractive idea would be that the ER interacts too strongly with the nonspecific DNA acceptor sites to allow it to find the specific DNA acceptor sites. In this

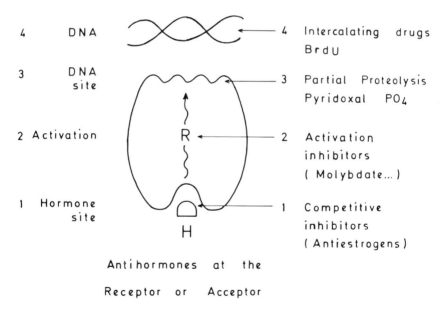

Figure 2. Inhibition of estrogen action at the receptor and acceptor level. Different classes of estrogen antagonists can prevent estrogen action at the receptor (R) level by inhibiting hormone binding (step 1) (classic antiestrogens), or the receptor-activation process (step 2), or the DNA, or nuclear-binding site of the receptor (step 3). Estrogen action can also be blocked at the DNA-acceptor level (step 4) after modification of DNA structure by BrdU substitution or DNA intercalation. The estrogen receptor is represented schematically as a whole complex with two binding sites (for the hormone and the DNA), without specifying whether these sites are located on one or different subunits or protein since the structure of the pure ER is not yet known.

respect, nt^i variants have been described in glucocorticoid-resistant systems in which the *in vitro* interaction of the receptor to DNA is increased compared to the interaction in glucocorticoid sensitive systems (Sibley and Tomkins, 1974).

ACKNOWLEDGMENTS.   This study was supported by INSERM, the NCI- INSERM cooperation on "Hormones and Cancer," the Foundation for French Medical Research, and an INSERM grant (No49-77-81). We thank members of the research Unit 148 of INSERM, particularly Miss E. Barrie for her help in the preparation of the manuscript. We are grateful to Drs. M. Lippman, M. Rich, and the Mason Research Laboratory (Dr. Bogden), for their gifts of human mammary cell lines.

REFERENCES

Adams, D. J., Edwards, D. P., and Mc Guire, W. L., 1980, Estrogen regulation of specific messenger RNA's in human breast cancer cells, *Biochem. Biophys. Res. Commun.* **97**:1354.

Adams, J., Garcia, M., and Rochefort, H., 1981, Estrogenic effects of physiological concentrations of 5-androstene-3β, 17β-diol and its metabolism in MCF₇ human breast cancer cells, *Cancer Res.* **41**:4720.

André, J., Vic, P., Humeau, C., and Rochefort, H., 1977, Nuclear translocation of the estradiol receptor: Partial inhibition by ethidium bromide, *Mol. Cell. Endocrinol.* **8**:225.

Blue, M. L., and Williams, D. L., 1981, Induction of avian serum apolipoprotein II and vitellogenin by tamoxifen, *Biochem. Biophys. Res. Commun.* **98**:785.

Borgna, J. L., and Rochefort, H., 1980, High-affinity binding to the estrogen receptor of ³H 4-hydroxytamoxifen, and active antiestrogen metabolite, *Mol. Cell. Endocrinol.* **20**:71.

Borgna, J. L., and Rochefort, H., 1981, Hydroxylated metabolites of tamoxifen are formed in vivo and bound to estrogen receptor in target, *J. Biol. Chem.* **256**:859.

Bouton, M. M., and Raynaud, J. P., 1978, The relevance of kinetic parameters in the determination of specific binding to the estrogen receptor, *J. Steroid Biochem.* **9**:9.

Capony, F., and Williams, D. L., 1981, Antiestrogen action in avian liver: The interaction of estrogens and antiestrogens in the regulation of apolipoprotein B synthesis, *Endocrinology* **108**:1862.

Chalbos, D., Vignon, F., and Rochefort, H., 1982, Effect of estrogen on cell proliferation and secretory proteins in the T47D human breast cancer cell line, *Cold Spring Harbor Conference on Cell Proliferation,* Vol. 9, p. 845, Cold Spring Harbor Laboratory, New York.

Clark, J., Baulieu, E. E., Baxter, J. D., de Crombrugghe, B., Jorgenson, E. C., Katzenellenbogen, B. S., Katzenellenbogen, J. A., Luebke, K., Moran, J., Rochefort, H., Sherman, M. R., and Töpert, M., 1976, in: Intracellular receptors group report, *Hormone and Antihormone Action at the Target Cell,* Dahlem Konferenzen, Life Science Research Report (J. H. Clark, W. Klee, A. Levitzki, and J. Wolff, eds.), pp. 147–169, Berlin.

Coezy, E., Borgna, J. L., and Rochefort, H., 1982, Tamoxifen and metabolites in MCF₇ cells: Binding to estrogen receptor and cell growth inhibition, *Cancer Res.* **41**:317.

Evans, E., Baskevitch, P. P., and Rochefort, H., 1982, Estrogen receptor DNA interaction: Difference between activation by estrogen and antiestrogen, *Eur. J. Biochem.* **128**:185.

Garcia, M., and Rochefort, H., 1977, Androgens on the estrogen receptor. II. Correlation between nuclear translocation and uterine protein synthesis, *Steroids* **29**:111.

Garcia, M., and Rochefort, H., 1978, Androgen effects mediated by the estrogen receptor in 7,12-dimethylbenz(a)anthracene-induced rat mammary tumors, *Cancer Res.* **38**:3922.

Garcia, M., Westley, B., and Rochefort, H., 1981, 5-Bromodeoxyuridine specifically inhibits the synthesis of estrogen-induced proteins in MCF$_7$ cells, *Eur. J. Biochem.* **116**:297.

Herschko, A., Mamont, P., Shields, R., and Tomkins, G. M., 1971, Pleiotypic response, *Nature New Biol.* **232**:206.

Horwitz, K. B., and Mc Guire, W. L., 1978, Estrogen control of progesterone receptor in human breast cancer, *J. Biol. Chem.* **253**:2223.

Kallos, J., Hollander, V. P., Fasy, T. M., and Bick, M. D., 1978, Estrogen receptor binds more tightly to bromodeoxyuridine-substituted DNA, *Proc. Natl. Acad. Sci. U.S.A.* **75**:4896.

Lippman, M., Bolan, G., and Huff, K., 1976, The effects of estrogens and antiestrogens on hormone responsive human breast cancer in long-term tissue culture, *Cancer Res.* **36**:4595.

McGuire, W. L., 1980, Steroid hormone receptor in breast cancer treatment strategy, in: *Recent Progress in Hormone Research* (R. O. Greep, ed.), Vol. 36, pp. 135-156, Academic Press, New York.

Notides, A., and Gorski, J., 1966, Estrogen-induced synthesis of a specific uterine protein, *Proc. Natl. Acad. Sci. U.S.A.* **56**:230.

O'Farrell, P. H., 1975, High resolution two-dimensional electrophoresis proteins, *J. Biol. Chem.* **10**:4007.

Palmiter, R. D., Mulvihill, E. R., Mc Knight, G. S., and Senear, A. W., 1978, Regulation of gene expression in the chick oviduct by steroid hormones, *Cold Spring Harbor Symp. Quant. Biol.* **42**:639–647.

Rochefort, H., and Borgna, J. L., 1981, Differences between estrogen receptor activation by estrogen and antiestrogen, *Nature* **292**:257.

Rochefort, H., and Capony, F., 1973, Etude comparée du comportement d'un antioestrogène et de l'oestradiol dans les cellules utérines, *C. R. Acad. Sci. Paris* **276**:2321.

Rochefort, H., Garcia, M., Vignon, F., and Westley, B., 1980, Proteins induced by the estrogen receptor in uterus and breast cancer cells, in: *Steroid Induced Uterine Proteins* (M Beato, ed.), pp. 171–181, Elsevier/North-Holland Biomedical Press, Amsterdam.

Rochefort, H., Capony, F., and Borgna, J. L., 1981a, Metabolism and binding of non-steroidal antiestrogen in mammals and chicken, in: *Non-Steroidal Antiestrogens* (R. L. Sutherland and V. C. Jordan, eds.), pp. 85–94, Harcourt Brace Jovanovitch Group (Australia), Academic Press, Sydney.

Rochefort, H., Coezy, E., Joly, E., Westley, B., and Vignon, F., 1981b, Hormonal control of breast cancer in cell culture, in: *Hormones and Cancer* (S. Iacobelli, R. J. B. King, H. R. Lindner, and M. E. Lippman, eds.), pp. 21–29, Raven Press, New York.

Ruh, T. S., and Ruh, M. F., 1975, Androgen induction of a specific uterine protein, *Endocrinology* **97**:1144.

Shyamala, G., and Leonard, L. J., 1980, Inhibition of uterine estrogen receptor transformation by sodium molybdate, *J. Biol. Chem.* **255**:6028.

Sibley, C. H., and Tomkins, G. M., 1974, Mechanism of steroid resistance, *Cell* **2**:221.

Sirbasku, D. A., and Benson, R. H., 1979, Estrogen-inducible growth factors that may act as mediators (estromedins) of estrogen-promoted tumor cell growth, in: *Hormones and Cell Culture* Cold Spring Harbor Conferences on Cell Proliferation (G. Sato and R. Ross, eds.), Vol. 6, p. 477–497, Cold Spring Harbor Laboratory, New York.

Sonnenschein, C., and Soto, A. M., 1980, But . . . Are estrogens *per se* growth-promoting hormones?, *J. Natl. Cancer Inst.* **64**:211.

Sutherland, R., Mester, J., and Baulieu, E. E., 1977, Tamoxifen is a potent "pure" antiestrogen in chick oviduct, *Nature* **267**:434.

Sutherland, R. L., Murphy, L. C., Foo, M. S., Green, M. D., and Whybourne, A. M., 1980, High-affinity antiestrogen binding site distinct from the estrogen receptor, *Nature* **288**:273.

Westley, B., and Rochefort, H., 1979, Estradiol induced proteins in the MCF₇ human breast cancer cell line, *Biochem. Biophys. Res. Commun.* **90:**410.

Westley, B., and Rochefort, H., 1980, A secreted glycoprotein induced by estrogen in human breast cancer cell lines, *Cell* **20:**352.

Westley, B., Vignon, F., Garcia, M., and Rochefort, H., (1980), A 46 K estrogen induced protein secreted by human breast cancer cells in culture. Second International Congress on Cell Biology, August 31–September 5, 1980, West Berlin.

Zava, D. T., and McGuire, W. L., 1978, Human breast cancer: Androgen action mediated by estrogen receptor, *Science* **199:**787.

3

The Estrogen-Induced/Dependent Renal Adenocarcinoma of the Syrian Hamster

Paul H. Naylor, Rabinder N. Kurl, Janet M. Loring, and Claude A. Villee

1. Introduction

The events that occur in a tissue prior to a detectable malignant transformation are complex, and may be unique for each type of tumor. Alternatively, a wide variety of neoplasias may have many preneoplastic events in common. The several hormone-dependent tumors make up one class of neoplasms which may have similar molecular events resulting in their induction and growth. Following long-term estrogen treatment the Syrian hamster develops renal adenocarcinomas which are both estrogen-induced and estrogen-dependent (Kirkman, 1959). We have been using this model to find answers for several questions concerning the induction and growth of hormone-dependent tumors with emphasis on the role of estrogen at the molecular level.

First described by Matthews *et al.* (1947), estrogen-induced/dependent kidney tumors in the hamster have many similarities with human kidney tumors (Horning and Whittick, 1954; Bloom *et al.*, 1963; Hogan, 1979). Both adenocarcinomas occur in the proximal convoluted tubules. Both often metastasize extensively. Both occur later in life and are predominantly found in males.

Paul H. Naylor, Rabinder N. Kurl, Janet M. Loring, and Claude A. Villee ● Department of Biological Chemistry and Laboratory of Human Reproduction and Reproductive Biology, Harvard Medical School, Boston, Massachusetts 02115.

Several excellent reviews of the tumor have been published, so only a brief review of the past with emphasis on the data pertinent to our developing hypotheses has been included in this paper (cf. Kirkman, 1959; Bloom *et al.*, 1963; Hogan, 1979).

Renal carcinogenesis occurs in more than 80% of our male hamsters by 9–12 months after the beginning of estrogen treatment. All biologically active estrogens, including the synthetic estrogen, diethylstilbestrol, induce the tumor whereas inactive steroids such as 17α-estradiol are not effective (Kirkman, 1959). Intact female hamsters fail to develop tumors unless castrated or neonatally masculinized by the injection of testosterone (Kirkman, 1959). Administration of progesterone or of antiestrogens along with the estrogen prevents tumor formation whereas castration of males has minimal effect on the induction of the tumor (Kirkman, 1959; Antonio *et al.*, 1974; Lin *et al.*, 1980a).

Initial investigation into the mechanism by which estrogens induce tumors of the kidney focused on defining the specific changes that occurred in the kidney as a result of administration of estrogen and on proving that such changes could be mediated by an estrogen receptor in the kidney. Early studies had demonstrated that hamster kidneys were presumably targets of estrogen since hyperplasia, accompanied by increases in enzyme levels, followed the administration of estrogens. Receptors for estrogens have been identified in the hamster kidney, whereas rats that fail to develop tumors in response to estrogen also appear to lack estrogen receptors in the kidneys (Bloom *et al.*,1963; Li *et al.*, 1975; Saluja *et al.*, 1979; Lin *et al.*, 1980a).

Both *in vivo* and *in vitro* experiments showed that the properties of the hamster kidney receptors are similar to those previously described for receptors in the uterus, in their affinity, specificity, and other characteristics as well as in their actions such as inducing an increase in progesterone receptors (Li *et al.*, 1974, 1976, 1977, 1978; DeVries *et al.*, 1972; King *et al.*, 1970).

While the above experiments clearly documented the importance of estrogen in inducing kidney tumors, estrogen administration also results in hyperplasia of the pituitary in both hamsters and rats (Vazquez-Lopez, 1944; Lloyd *et al.*, 1973). Vazquez-Lopez first reported (1944) that male and female hamsters with pellets implanted containing 10 mg of either estradiol benzoate or diethylstilbestrol had enlarged pituitaries after 270 days of treatment. This observation has been confirmed by various laboratories (Koneff *et al.*, 1946; Kirkman, 1950; Horning and Whittick, 1954; Hamilton *et al.*, 1975; Lin *et al.*, 1982). Russfield (1963) reported that the pituitary adenomas produced by diethylstilbestrol in male hamsters were of the intermediate lobe type in castrated males and of the chromophobe type in intact males. Using ultrastructural techniques Hamilton *et al.* (1975) confirmed the findings of Russfield (1963). It was reported that the intermediate lobe had undergone gross enlargement, with hyperplasia and invasion of the posterior lobe, infundibular stalk, and, in some animals, the anterior lobe (Hamilton *et al.*, 1975). We have also observed that cells of the intermediate lobe are enlarged and irregular in shape as compared with those in hamsters not treated with estrogen. Ultrastructural examination revealed that the cells of the intermediate lobe were secretory and that diethylstilbestrol significantly increased the number of prolactin-secreting cells and decreased the number of somatotropin-secreting cells and basophils in the anterior lobe of the pituitary gland.

Hamilton *et al.* (1975) treated hamsters with a prolactin-inhibiting drug, 2-bromo-ergocryptine methane sulfonate (CB 154) along with diethylstilbestrol. Treating the animals with the prolactin-inhibiting drug decreased the weight of the pituitary gland significantly compared to animals treated with diethylstilbestrol alone. However, the glands had significantly greater weights than those of untreated control animals. There was a significant reduction in the number of prolactin-secreting cells and an increase in the number of cells secreting growth hormone. CB 154 administered concurrently with diethylstilbestrol decreased the incidence and severity of renal tumors but CB 154 failed to produce regression of tumors that were already established. These findings suggest that the pituitary may play a permissive role in the genesis of kidney lesions. The results were in conflict with studies reported by Kirkman (1959) that tumors can be induced in hypophysectomized hamsters. The work by Kirkman (1959) is questionable in many respects. For example, in one experiment four animals were hypophysectomized when they were between 357 and 388 days old and after the diethylstilbestrol pellets had been removed for 2–6 weeks. The time period for which diethylstilbestrol pellets were administered; 309–344 days, is one in which most hamsters have already developed lesions. Hence, little if any conclusion can be drawn on the effect of hypophysectomy in the animals at that time. In all of the experiments the very small number of animals used and the lack of validation of complete hypophysectomy of the animals brought the results into question (Hamilton *et al.*, 1975; Lin *et al.*, 1982).

The importance of the pituitary in renal tumorigenesis is further suggested by the observation that female hamsters which have been neonatally masculinized with androgens develop renal tumors after long-term estrogen treatment whereas nonmasculinized females do not (Kirkman, 1959). It has been postulated that high levels of progesterone secreted by the corpora lutea could inhibit tumorigenesis just as progesterone administered simultaneously with estradiol inhibits the development of renal tumors (Kirkman, 1959). It is questionable, however, that chronic estrogen treatment results in sufficiently high levels of progesterone to antagonize estrogen-induced tumorigenesis since such a treatment should lead to acyclic females. Thus, an effect at the pituitary level rather than the ovary may be more critical to the prevention of tumor formation in females.

2. Hypophysectomized Hamsters

The importance of the pituitary in the estrogen induction of neoplasia has been controversial, mainly as a result of the early report of Kirkman (1959) that hypophysectomized animals developed renal tumors (see Section 1). Since the

Table I. The Effect of Hypophysectomy and Length of Diethylstilbestrol (DES)
Treatment

Treatment protocol	Number of animals	Number with tumor	Tumor incidence (%)
Castrate (DES for 9–12 months)	20	16	80
Castrate, hypophysectomized (DES for 9–12 months)	20	0	0
Castrate (DES for 1–2 months before transplantation of tumor tissue)	10	8	80
Castrate, hypophysectomized (DES for 1–2 months before transplantation of tumor tissue)	12	0	0
Castrate (DES for 1–2 weeks before transplantation of tumor tissue)	8	2	25
Castrate (DES for 3–4 weeks before transplantation of tumor tissue)	8	7	87

examination of Kirkman's data showed them to be less than convincing, we undertook to repeat the experiment and found that estrogen implants do not induce renal tumors in hypophysectomized hamsters (Lin *et al.*, 1982). Both transplanted tumors and primary tumors fail to develop in hypophysectomized hamsters implanted with estrogen pellets. The levels of follicle-stimulating hormone, luteinizing hormone, and prolactin in the serum were also reduced to minimally detectable levels, indicating complete hypophysectomy (Lin *et al.*, 1982) (Table I). Attempts to induce tumors by administering prolactin (Hamilton *et al.*, 1979), to increase the incidence by simultaneously giving prolactin and diethylstilbestrol to intact hamsters (Hamilton *et al.*, 1975), *or* in hypophysectomized hamsters by administration of estrogen (Lin *et al.*, 1982) have not been successful. This may be accounted for by the recent reports that prolactin should be administered in polyvinylpyrrolidone in order to maintain a significant increase in serum levels. Alternatively, a combination of known, or as yet unidentified, pituitary products may act as cocarcinogens, promoters, or even direct tumor inducers.

3. Pituitary Cells in Culture

It has been well established that the hypothalamus exerts a primary control over the production and release of gonadotropins by the pituitary. In addition, the estrogens have a direct effect on pituitary function (King and Mainwaring, 1974; Arafah *et al.*, 1980). Because of the complexities encountered while studying pituitary secretion *in vivo* we cultured hypophyses from normal animals,

Table II. α-MSH and Prolactin Levels in Pituitary
Culture Media

Medium	α-MSH (pg/ml)	Prolactin (ng/ml)
Hyperplastic pituitary		
Control	152 ± 20	1.07 ± 0.29
Day 4	1957 ± 74	11.03 ± 0.93
Day 7	1136 ± 89	11.14 ± 1.18
Normal pituitary		
Control	142 ± 8	1.07 ± 0.29
Day 4	161 ± 10	5.05 ± 0.50
DES-treated pituitary[a]		
Day 4	244 ± 8.5	5.70 ± 0.15

[a] Animals treated with diethylstilbestrol for 2 months.

animals treated with diethylstilbestrol, and tumor-bearing animals. We then assayed the culture media in which the pituitary cells had been growing for alpha-melanocyte-stimulating hormone (α-MSH) and for prolactin. Cell cultures of hyperplastic pituitaries from animals treated with diethylstilbestrol secreted greater amounts of α-MSH and prolactin as measured by radioimmunoassay than did cultures of normal pituitaries (Table II). The maximal concentration of these polypeptides in the medium was achieved 4–7 days after initiating cell culture. These high concentrations were maintained in the media for as long as 10 days.

Normal pituitaries maintained in cell culture secrete primarily prolactin whereas the amount of α-MSH was barely detectable. Similar results were obtained when pituitary glands from animals treated with diethylstilbestrol for 2 months were cultured. The addition of diethylstilbestrol to the medium did not stimulate the secretion of prolactin. Using a GH cell line Dannies *et al.* (1976) were unable to show any stimulating effect of estrogens on prolactin secretion, though this cell line was initially reported to respond to estrogens. Whether the hyperplastic pituitaries secrete some substance into the media which enhances tumorigenesis or tumor growth is not yet known. Based on the evidence to date either prolactin or α-MSH could be candidates for such an activity.

4. Renal Adenocarcinoma Cells in Culture

The difficulties inherent in studying tumor growth *in vivo* led us to develop methods for growing renal adenocarcinoma cells in primary culture. Tumor cells grow well in Dulbecco's modified minimal essential medium supplemented with 10% calf serum and can be passaged two or three times before the growth rate declines significantly. Estrogens are required to maintain the concentrations of

TABLE III. Effect of DES Concentration on Cell Growth and DNA Synthetic Activity of Renal Adenocarcinoma[a]

DES concentration (M)	DMEM[b] + 10% calf serum		DMEM + 10% fetal calf serum		DMEM + 5%/5%[c]	
	(cells/ml)	(cpm/800 μl)	(cells/ml)	(cpm/800 μl)	(cells/ml)	(cpm/800 μl)
0	14.4 ± 0.4	4144 ± 393	29.6 ± 8.2	10,158 ± 2148	24.7 ± 1.5	7,847 ± 966
10^{-7}	19.3 ± 1.8	2660 ± 440	32.4 ± 3.0	10,443 ± 2117	29.2 ± 0.1	12,035 ± 465
10^{-8}	17.9 ± 1.5	1654 ± 237	38.4 ± 3.8	11,125 ± 1770	27.6 ± 2.5	11,743 ± 401
10^{-9}	12.2 ± 1.1	1185 ± 193	38.0 ± 3.8	13,924 ± 2518	34.7 ± 0.5	13,703 ± 2917

[a] Values assessed at day 5 after primary culture and expressed as mean ± SEM for two to four wells
[b] DMEM is Dulbecco's minimum essential medium.
[c] 5%/5%, 5% fetal calf serum + 5% calf-serum-supplemented media

estrogen receptors and progesterone receptors to the level initially present in the cells in culture (Lin *et al.*, 1978, 1980a). These findings are consistent with our evidence that the progesterone receptor is elevated in both tumor tissue and kidney from hamsters treated with diethylstilbestrol when compared with animals not treated with the estrogen (Lin *et al.*, 1977). Diethylstilbestrol does not significantly elevate the activity of ornithine decarboxylase in either cultured normal cells or cultured tumor cells, although ornithine decarboxylase activity is elevated in tumor tissue compared to the surrounding normal kidney tissue (Naylor *et al.*, submitted). We interpret this to mean that the activity of ornithine decarboxylase in the kidney may not be directly controlled by estrogens *in vivo* whereas the amount of progesterone receptors in the kidney is directly controlled by estrogen. It was clear, however, that both normal cells and kidney tumor cells grown in media fortified with calf serum had greater ornithine decarboxylase activities and higher rates of cell growth compared to cells cultured for 24 hr without calf serum. Under no experimental conditions could we consistently demonstrate an increased rate of cell growth or increased rate of synthesis of DNA in cells exposed to diethylstilbestrol compared to controls not exposed to estrogen (Table III). Because of our results demonstrating elevated concentrations of αMSH and prolactin in serum from hamsters with tumors (see Section 5) we tested various combinations of diethylstilbestrol, α-MSH, and prolactin on cell growth in culture (Table IV). To date we have not been successful in either stimulating the synthesis of DNA or increasing the rate of cell growth with any combination of these hormones.

The unexpected results from these studies made it imperative to assess the growth properties of hamster serum from animals treated under various protocols. Our initial objective was to determine why renal adenocarcinoma cells transplanted into the hamster would grow *in vivo* only if diethylstilbestrol was present for

Table IV. Effect of Hormones on DNA Synthetic Activity of Renal Adenocarcinoma Cells in Culture

Hormone (10^{-8}M)	Media without DES[a] (cpm/800 μl)	Media with DES (10^{-8})[a] (cpm/800 μl)
Saline	65,530	57,184
Diethylstilbestrol (DES)	73,445	59,606
Prolactin (PRL)	66,983	48,056
α-MSH	62,148	49,962
DES + PRL	84,804	64,386
DES + α-MSH	72,055	69,759
DES + PRL + α-MSH	65,046	56,715
PRL + α-MSH	61,464	49,960

[a] Assayed at day 3 after primary culture.

Table V. Effect of Hamster Serum on Renal Adenocarcinoma Cell Growth

	Days 3–4		Days 5–6	
Experiment	Cells ($\times 10^4$/ml)	DNA Synthesis (cpm/ml)	Cells ($\times 10^4$/ml)	DNA Synthesis (cpm/ml)
1 (5% CS)				
NHS	6.0	1,532	6.3	742
HTS	8.0	29,294	17.8	48,866
FCS	5.6	11,890	10.9	11,175
2 (5% CS)				
NHS	13.7	7,542	11.7	11,626
HTS	12.7	14,074	12.4	21,038
FCS	10.3	16,975	10.8	24,369
3 (5% CS)				
NHS	8.5	40,530	13.6	46,215
HTS	10.3	56,477	15.7	62,282
FCS	10.5	39,700	15.6	65,730
4 (5% FCS)				
NHS	11.9	9,980	9.6	9,706
NTS	12.5	13,761	13.2	28,706

[a] Either calf serum (CS) or fetal calf serum (FCS) was added at 5% to all cultures in addition to 5% of the sera mentioned (NHS, normal hamster serum; HTS, hamster tumor serum).

the previous 2–4 weeks, but would grow very well *in vitro* whether or not diethylstilbestrol was present. The results to date are complex and complicated by variations not only in the rates of tumor growth in culture but also in the degree of activity of the serum from different animals.

The protocol is to add hamster serum (5%) to calf serum (5%) supplemented Dulbecco's minimal essential medium and to assess cell growth and the rate of DNA synthesis over days 3–7, the days of greatest cell growth. All assays were carried out in triplicate or quadruplicate and the variations between and within assays were less than 20%. We have found consistently that fetal calf serum stimulates growth better than calf serum. The response to serum from both tumor-bearing hamsters and control hamsters was more variable (Table V). In all cases the tumors are dependent upon diethylstilbestrol, but their source varies. The term "primary" indicates the cells were from tumors that were induced by diethylstilbestrol and obtained from diethylstilbestrol-treated hamsters whereas transplanted cells have been given to diethylstilbestrol-primed hamsters in a variety of sites. To date, the hamster serum is consistently active at lower dilutions than fetal calf serum. Experiments are now in progress to quantitate the activity and to determine its source in the serum.

5. Hormone Levels in Hamster Serum

In response to long-term estrogen treatment there is a concomitant increase in the number of prolactin-secreting cells of the anterior pituitary and of Type I light cells of the intermediate lobe of the pituitary gland believed to be the source of α-MSH (Hamilton *et al.*, 1975). This prompted us to study the profile of these hormones in the serum of hamsters treated either acutely or chronically with diethylstilbestrol. There is an initial increase in the concentration of α-MSH in the serum 1 week after treatment with diethylstilbestrol. A further increase is observed 4-5 months after the treatment is begun and thereafter the concentration of α-MSH in the serum remains very high. Very high concentrations of α-MSH were found in the serum of tumor-bearing animals. The concentration of prolactin in the serum increases steadily, reaching a peak at 4 months after the treatment with diethylstilbestrol was begun, but thereafter the levels decline over a further period of 4–6 months (Table VI). This results in concentrations of prolactin in the serum of tumor-bearing animals which are not different from that of control animals. The relatively low concentrations of prolactin in the serum of tumor-bearing animals may explain why the administration of CB154 failed to cause regression of the renal tumor (Hamilton *et al.*, 1975). Measurements of the concentration of α-MSH in serum were reported by Saluja *et al.* (1979). They did not report an increase in the concentration of α-MSH in the serum of hamsters 1 week after the initiation of treatment with diethylstilbestrol. This may be due to our using castrate males whereas Saluja *et al.* used intact males. We have observed that castration lowers the concentration of α-MSH in the serum and, after 1 week of treatment with diethylstilbestrol, α-MSH returns to the levels observed in intact animals.

Table VI. α-MSH and Prolactin Levels in
Hamster Serum at Various Times after
Diethylstilbestrol (DES)

Time with DES (weeks)	α-MSH[a] (pg/ml)	Prolactin[a] (ng/ml)
0	587 ± 80	5.4 ± 1.1
1	976 ± 85	11.4 ± 1.5
9	1,064 ± 180	11.3 ± 2.9
14	1,064 ± 53	19.8 ± 5.0
20	2,954 ± 268	7.3 ± 1.7
>40	>3,383	7.8 ± 0.4

[a] Values are mean ± SEM for three to six animals/group.

In contrast to the increase in α-MSH and prolactin in the serum, the concentrations of luteinizing hormone and follicle-stimulating hormone in the serum of animals treated with diethylstilbestrol decreased significantly (Lin *et al.*, 1982). The two gonadotropins decrease gradually and the tumor-bearing animals have barely detectable levels of these hormones. These observations are in agreement with the histological findings of Hamilton *et al.* (1975) in which there was a decrease in the chromophobe type cells.

6. Tumorogenesis via Immunoendocrine Perturbations

The experiments described above have suggested that the mechanism of tumor induction may be more complex than originally postulated. It seems obvious that the administration of estrogen causes many complex changes in the endocrine and immune status of the hamster. The events involved in the induction of the tumor and those involved in its growth may not be the same. The role of the pituitary in this process is undeniable but is not yet completely defined. The possibility that diethylstilbestrol elevates the concentration of some tumor growth factor in the serum is becoming more likely. Determinations of whether this growth-promoting activity emanates from the pituitary, the tumor, or from some other as yet unidentified site, and whether it is universal in nature, must wait until the factor has been identified.

The experiments with cell cultures of renal adenocarcinoma will be a key tool in our efforts to clarify the molecular events of tumor growth. Further experiments with pituitary cells in culture should provide important clues regarding the nature of any growth-promoting as well as tumor-inducing activity. These could be tested both on renal adenocarcinoma and normal cells in culture. We also are attempting to induce tumors using pellets of diethylstilbestrol in hypophysectomized hamsters in which pituitary tissue has been transplanted into the cheek pouch. Finally we plan to define the effects of the administration of estrogen on the immune status of the animal. It is to be hoped that integration of the results of these several types of experiments will provide the first steps to understanding the molecular events which result in renal tumorigenesis via estrogen administration to the hamster.

ACKNOWLEDGMENT. Supported by Grant No. CA 24,615 from the National Cancer Institute, National Institutes of Health.

REFERENCES

Antonio, P., Gabaldon, M., Lacomba, T., and Juan, A., 1974, Effect of the antiestrogen nafoxidine on the occurrence of estrogen-dependent renal tumors in hamsters, *Horm. Metabol. Res.* **6:**522.

Arafah, B. M., Manni, A., and Pearson, O. H., 1980, Effect of hypophysectomy and hormone replacement on hormone receptor levels and the growth of 7,12-dimethylbenz(a)anthracene-induced mammary tumors in the rat, *Endocrinology* **107**:1364.

Bloom, H. J. G., Dukes, C. E., and Mitchley, B. G. V., 1963, The estrogen-induced renal tumor of the Syrian hamster. Hormone treatment and possible relationship to carcinoma of the kidney in man, *Br. J. Cancer* **17**:611.

Dannies, P. S., Gautvik, R. M., and Tashjian, A. H., Jr., 1976, A possible role of cyclic AMP in mediating the effects of thyrotropin-releasing hormone in prolactin release and in prolactin and growth hormone synthesis in pituitary cells in culture, *Endocrinology* **98**:1147.

DeVries, J. R., Ludens, J. H., and Fanestil, D. D., 1972, Estrogen renal receptor molecules and estradiol-dependent antinatiuresis, *Kidney Int.* **2**:95.

Hamilton, J. M., Flaks, A., Saluja, P. G., and Maguire, S., 1975, Hormonally induced renal neoplasia in the male Syrian hamster and the inhibitory effect of 2-bromo-α-ergocryptine methanesulfate, *J. Natl. Cancer Inst.* **54**:1385.

Hamilton, J. M. Saluja, P. G., Thody, A. J., and Flask, A., 1979, The pars intermedia and renal carcinogenesis in hamsters, *Eur. J. Cancer* **13**:29.

Hogan, T. F., 1979, Hormonal therapy of renal-cell carcinoma, in: *Endocrinology of Cancer*, Vol. II (D. P. Rose, ed.), p. 69, CRC Press, Bogata, Florida.

Horning, E. S., and Whittick, S. W., 1954, The histogenesis of stilboestrol-induced renal tumor in the male Golden hamster, *Br. J. Cancer* **8**:451.

King, R. J. B., Smith, J. A., and Steggles, A. W., 1970, Estrogen binding and the hormone responsiveness of tumors, *Steroidologia* **1**:73.

King, R. J. B., and Mainwaring, W. I. P., 1974, *Steroid–Cell Interactions*, Butterworths, London.

Kirkman, H., 1950, Different types of tumors observed in treated and in untreated golden hamsters, *Anat. Rec.* **106**:277.

Kirkman, H., 1959, Estrogen-induced tumors of the kidney in Syrian hamsters, *Natl. Cancer Inst. Monogr.* **1**:1.

Koneff, A. A., Simpson, M. E., and Evans, H. M., 1946, Effects of chronic administration of diethylstilbestrol on the pituitary and other endocrine organs of the hamster, *Anat. Rec.* **94**:169.

Li, J. J., Talley, D. J., Li, S. A., and Villee, C. A., 1974, An estrogen binding protein in the renal cytosol of intact, castrated and estrogenized Golden hamsters, *Endocrinology* **95**:1134.

Li, J. J., Li, S. A., Klein, L. A., and Villee, C. A., 1975, Dehydrogenase isozymes in the hamster and human renal adenocarcinoma, in: *Isozyme*, Vol. III (C. L. Markert, ed.) p. 837, Academic Press, New York.

Li, J. J., Talley, D. J., Li, S. A., and Villee, C. A., 1976, Receptor characteristics of specific estrogen binding in the renal adenocarcinoma of the Golden hamster, *Cancer Res.* **36**:1127.

Li, S. A., Li, J. J., and Villee, C. A., 1977, Significance of the progesterone receptor in the estrogen induced and dependent renal tumor of the Syrian hamster, *Ann. N.Y. Acad. Sci.* **286**:369.

Li, J. J., Cuthbertson, T. L., and Li, S. A., 1980, Inhibition of estrogen tumorigenesis in the Syrian golden hamster kidney by antiestrogens, *J. Natl. Cancer Inst.* **64**:795.

Lin, Y. C., Talley, D. J., and Villee, C. A., 1978, Progesterone receptor levels in estrogen-induced renal carcinomas after serial passage beneath the renal capsule of Syrian hamsters, *Cancer Res.* **28**:1286.

Lin, Y. C., Loring, J. M., and Villee, C. A., 1980a, Diethylstilbestrol stimulates ornithine decarboxylase in kidney cells in culture, *Biochem. Biophys. Res. Commun.* **95**:1393.

Lin, Y. C., Talley, D. J., and Villee, C. A., 1980b, Dynamics of progesterone binding in nuclei and cytosol of estrogen-induced adenocarcinoma cells in primary culture, *J. Steroid Biochem.* **13**:29.

Lin, Y. C., Loring, J. M., and Villee, C. A., 1982, A permissive role of the pituitary in the induction and growth of estrogen-dependent renal tumors, *Cancer Res.* **42**:1015.

Lloyd, H. M., Meares, J. D., and Jacobi, J., 1973, Early effects of stilbestrol on growth hormone and prolactin secretion and on pituitary mitotic activity in the male rat, *J. Endocrinol.* **58:**227.

Matthews, V. S., Kirkman, H., and Bacon, R. L., 1947, Kidney damage in the Golden hamster following chronic administration of diethylstilbestrol and sesame oil, *Proc. Soc. Exp. Med.* **66:**195.

Russfield, A. B., 1963, Effect of castration on induction of pituitary tumors with diethylstilbestrol in the Syrian hamster, *Proc. Am. Assoc. Cancer Res.* **4:**59.

Saluja, P. G., Hamilton, J. M., Thody, A. J., Isnail, A. A., and Knowles, J., 1979, Ultrastructure of intermediate lobe of the pituitary and melanocyte-stimulating hormone secretion in oestrogen-induced kidney tumors in male hamsters, *Arch. Toxicol. Suppl.* **2:**41.

Vazquez-Lopez, E., 1944, The reaction of the pituitary gland and related hypothalamic centres in the hamster to prolonged treatment with estrogens, *J. Pathol. Bacteriol.* **65:**1.

4

Stimulation of Milk-Fat Synthesis in Mammary Epithelioid Cells by Progesterone

Robert T. Chatterton, Jr., Sheila M. Judge, Eldon D. Schriock, David Olive, John N. Haan, and Harold G. Verhage

1. Introduction

1.1. Specific Features of Carbohydrate Metabolism and Lipogenesis in Mammary Epithelium

While the mammary epithelium has a highly efficient mechanism for conversion of glucose to fatty acids, the metabolism of glucose is distinctly different from that occurring in adipose tissue. The pentose phosphate pathway predominates and glucose-6-phosphate dehydrogenase is much more sensitive to insulin (Leader and Barry, 1969; Oka *et al.*, 1974). In adipose tissue glucose, lactate, and pyruvate serve equally well as precursors of fatty acids, but in rat mammary tissue lipogenesis from pyruvate is much less than that from glucose, and pyruvate nearly completely inhibits utilization of glucose for fatty acid synthesis (Katz *et al.*, 1974). Also, the ratio of citrate cleavage enzyme to pyruvate carboxylase

Robert T. Chatterton, Jr., Sheila M. Judge, Eldon D. Schriock, David Olive, and John N. Haan • Department of Obstetrics and Gynecology, Northwestern University Medical School, Chicago, Illinois 60611   *Harold G. Verhage* • Department of Obstetrics and Gynecology, University of Illinois at the Medical Center, Chicago, Illinois 60680.

is very different, with mammary tissue having proportionally about five times more of the former. Cortisol acts particularly in the presence of some epinephrine to stimulate lipase activity and to inhibit transport of glucose in adipose tissue (Olefsky, 1975), but cortisol is required for lactation (Chatterton et al., 1979) and together with prolactin synergizes with insulin to stimulate lipogenesis in mammary tissue (Cameron et al., 1975). However, no direct evidence is available on the effect of cortisol on triglyceride turnover or lipase activity in mammary tissue.

The fatty acids themselves are different in mammary epithelium. In the presence of prolactin (Hallowes et al., 1973; Dils et al., 1972); fatty acid synthesis is terminated before chain elongation reaches C_{16}. The short- and medium-chain fatty acids are esterified at the 3-sn position of milk triglycerides.

1.2. Role of Insulin in Uptake and Oxidation of Glucose in Mammary Epithelium

In the lactating diabetic woman the requirement for insulin is decreased (Lawrence, 1980) apparently because of transfer of glucose from the blood to the breast for conversion to lactose. This observation may be indicative of glucose uptake by the breast that is relatively independent of insulin. Indeed, Robinson and Williamson (1977) found that lactating rats made diabetic by streptozotocin treatment did not have decreased uptake of glucose into mammary tissue, nor did similar animals exhibit a deficiency in lactose synthesis (Kyriakou and Kuhn, 1973). In vitro, insulin did not increase glucose uptake when glucose was present in a normal concentration (5 mM) (Williamson et al., 1974; Moretti and Abraham, 1966). However, insulin increases $^{14}CO_2$ production from [^{14}C]glucose (Moretti and Abraham, 1966) and was found by Nikolaeva and Balmukhanov (1980) to increase oxygen uptake by lactating rat mammary tissue by 40%. Oxidation of glucose is primarily through glycolysis and the pentose phosphate pathway (Smith and Abraham, 1975). Katz et al. (1974) found that 75–100% of NADPH in the mammary gland is supplied by the pentose cycle. In mammary tissue the cytosolic-reducing equivalents are considered to exceed utilization, and transport of the excess into mitochondria is likely (Katz et al., 1974).

1.3. Insulin Requirement for Biosynthesis of Milk Products

Norgren (1967) found that insulin was not required for growth and differentiation of the rodent mammary gland in vitro, but diabetes produced by alloxan treatment of lactating rats results in an immediate depression in lactational performance (Martin and Baldwin, 1971); synthesis of lipid as well as casein and lactose was decreased. Acute insulin insufficiency was expressed in mammary tissue most prominently as a failure of oxidation of reduced nicotinamide adenine dinucleotides (Martin and Baldwin, 1971). Failure to synthesize lipid in the absence of insulin clearly is not because of a deficiency of glucose or NADPH.

Insulin has been shown to stimulate an increase in immunoassayable fatty acid synthetase, but the activity of the enzyme was not rate-limiting under any conditions studied (Speake *et al.*, 1975). The most likely candidate for the rate-limiting step in lipogenesis is acetyl CoA carboxylase, which has the lowest activity of all enzymes involved in conversion of glucose to lipid (McLean *et al.*, 1972).

1.4. Interaction of Substrates in Utilization of Glucose for Lipid Synthesis

The presence of fatty acids in the medium provides a competing source of substrate for fat synthesis. Bailey *et al.* (1973) found that free fatty acids were extensively utilized by fibroblasts *in vitro*. Removal of fatty acids results in a rapid increase in incorporation of [^{14}C]glucose into fatty acids (Raff, 1970) and a four- to sixfold increase in acetyl CoA carboxylase activity (Jacobs *et al.*, 1973). In fact, the activation of fatty acid synthesis by insulin has been attributed to removal of acyl CoA, a potent inhibitor of acetyl CoA carboxylase (Halestrap and Denton, 1973). Similar studies in mammary tissue have not been done.

1.5. Role of Steroids in Mammary Lipogenesis

Steroids have important regulatory effects on lipogenesis in the mammary gland. Cameron *et al.* (1974) and Hallowes *et al.* (1973) have shown that fatty acid synthesis is increased *in vitro* when prolactin and cortisol are added to a medium containing insulin. Neither cortisol nor prolactin was effective when added alone with insulin. However, the significance of the cortisol effect *in vivo* is in doubt since adrenalectomy of rats resulted in no change in pathways of lipogenesis (Greenbaum and Darby, 1964). Chatterton *et al.* (1979) also found that ovariectomy of the midpregnant rat led to increases in prolactin levels and mammary secretory activity; simultaneous adrenalectomy suppressed the formation of lactose and casein, but fat synthesis continued as evidenced by the formation of extremely large fat droplets within the mammary epithelial cells. However, when prolactin secretion was suppressed in the ovariectomized rats by administration of 2-bromo-α-ergocriptine, no accumulation of lipid droplets occurred. Thus, there is considerable evidence that milk-fat biosynthesis that occurs in rodents is dependent on insulin and prolactin, but the importance of adrenocorticoids is not clearly established.

Progesterone can prevent the initiation of milk secretion at parturition but, when given to rats during lactation, it does not inhibit milk secretion (Walker and Matthews, 1949). Progestins given after lactogenesis may promote lactation in rodents, possibly by increasing insulin secretion (Sutter-Dub *et al.*, 1974). Progesterone, however, has also been shown to have a stimulatory effect on milk secretion by mammary tissue of the dog *in vitro* (Barnawell, 1967). The bulk of the evidence from studies of women receiving injectable progestins for contraception during lactation is that, if started after the first week after delivery, progestins do not inhibit established lactation, and frequently have been found to increase milk production (Hull, 1981).

1.6. Mammary Tumor Cells in Culture as a Model for Mammary Lipogenesis

Many studies have been carried out using mouse or rat mammary explants to study the effect of hormones on lipogenesis. Such explants contain abundant fat cells which do not utilize substrates or respond to hormones in the same way as the secretory cells of the mammary gland. Diffusion of substrates and hormones to the inner cell mass is a limiting factor that does not apply to the tissue *in vivo*. Dissociated cells from mammary glands offer some advantages in this regard, but since recovery of epithelial cells through the isolation procedure may be less than 50%, some selection among epithelial cells is certain to occur. Also, the use of collagenase and trypsin for dissociating the mammary cells may alter significantly the plasma membranes of the mammary cells, modifying the response to some hormones. In addition, the relatively short life span of explants or dissociated cells in culture means that the responses of these cells are constantly changing because of continuing cell death.

While normal cells in culture provide a means of assessing certain biochemical events that cannot be observed in any other way, transformed cells, if they maintain the biochemical pathways and response systems of interest, offer the advantage of a constantly viable, uniform source of cells.

We have investigated several human breast cancer cell lines for their potential as models for study of hormone induction of milk-product biosynthesis. One cell line, T-47D, isolated from a pleural effusion of a mammary ductal carcinoma (Keydar *et al.*, 1979), synthesizes triglycerides in culture. Based on the presence of short-chain fatty acids, suppression of synthesis by pyruvate, and ultrastructural appearance and localization of the fat droplets, it appears that the synthesis of triglycerides in culture is a vestige of normal mammary cell function retained by this tumor cell line. In keeping with evidence from studies of rodent mammary glands, adrenocortical steroids have little effect on triglyceride biosynthesis in these cells. However, progestins stimulate the accumulation of triglycerides by these cells, probably by decreasing lipase activity.

2. Materials and Methods

2.1. Materials

Solvents and reagents were reagent grade and were used without further purification. Lipid standards for thin-layer chromatography were obtained from Supelco, Bellefonte, PA. Steroids used in the study with their respective suppliers were 4-pregnene-3,20-dione (progesterone); 11α-hydroxy-4-pregnene-3,20-dione (11α-hydroxyprogesterone); 11β,17,21-trihydroxy-4-pregnene-3,20-dione (cortisol) and its acetate, 11β, 21-dihydroxy-4-pregnene-3,20-dione 21-acetate (corticosterone acetate); 20α-hydroxy-4-pregnen-3-one (20α-dihydroprogesterone); 20β-hydroxy-4-pregnen-3-one (20β-dihydroprogesterone); 9α-fluoro-11β,17,21-trihydroxy-16β-methyl-1,4-pregnadiene-3,20-dione (dexamethasone); 19-nor-17α-ethynyl-17β-hydroxy-4-androsten-3-one (norethindrone); 18-formyl-11β,21-dihydroxy-4-pregnene-3,20-dione (aldosterone) from Sigma Chemical Co., St. Louis, MO; 21-hydroxy-4-pregnen-3,20-dione (deoxycorticosterone); 17-hydroxy-4-pregnene-3-,20-dione (17-hydroxyprogesterone); 17β-hydroxy-5α-androstan-3-one (dihydrotestosterone), from Mann Research Laboratories, NY, NY; 17α,21-dimethyl-19-nor-pregna-4,9-diene-3,20-dione (R5020) from New England Nuclear Corp., Boston, MA; and 6-chloro-17-acetoxy-1α,2α-methylenepregna-4,6-diene-3,20-dione (cyproterone acetate) from Schering AG, Berlin, West Germany. Ovine prolactin (PRL) NIH-P-S10, was a gift from the NIAMDD, National Pituitary Agency, Baltimore, MD. Insulin was Lente U-100 from Eli Lilly and Co., Indianapolis, IN. L-Triiodothyronine (T_3) was from Sigma Chemical Co., St. Louis, MO, as was Norit A, dextran (mol. wt. 83,000), tris(hydroxyethylaminoethane), and Folin reagent. Rhodamine 6G, phenylhydrazine, and diphenylamine were from Eastman Organic Chemicals, Rochester, NY. Periodic acid was from G. Fredrick Smith Chemical Co., Columbus, OH. Silica gel G was from EM Labs, Elmsford, NY. Calf thymus DNA was from Calbiochem, San Diego, CA. [9,10-^{3}H(N)]-palmitate (SA, 12 Ci/mmole) was obtained from New England Nuclear Corp. [2-^{14}C]acetate, sodium salt, 58 mCi/mmole, was obtained from Amersham, Arlington Heights, IL. The liquid scintillation fluid, Budget Solve™, Research Products International Corp., Mt. Prospect, IL, was used for counting ^{3}H and ^{14}C.

Supplies for tissue culture including Eagle's modified minimum essential medium with Earle's salts (MEM), L-glutamine, sodium pyruvate, MEM nonessential amino acids, Earle's balanced salt solution without calcium or magnesium (EBSS), 25 mg/ml solution of crystalline trypsin, and fetal calf serum were obtained from Grand Island Biological Co. (GIBCO), Grand Island, NY. Tissue culture flasks with 25-cm^2 and 75-cm^2 surfaces were obtained from Corning Glass Works, Corning, NY.

2.2. Cell Culture

T-47D cells, passage number 74, were supplied by the E. G. and G Mason Research Institute (Rockville, MD). The cells were grown in open 75-cm^2 culture flasks in a water-jacketed incubator at 37°C in a humidified atmosphere of 95% air, 5% CO_2. Cultures were fed Mondays, Wednesdays, and Fridays by replacement of the 10 ml of medium. The attached cells were washed with 5 ml of growth medium before replacement on Mondays and Fridays. The growth medium consisted of MEM containing 10% by volume of fetal calf serum, 0.1 mM MEM nonessential amino acids, 1 mM pyruvate, and 0.2 U/ml of Lente insulin. Antibiotics were not included in the culture medium and all manipulations were carried out in a sterile laminar-flow hood.

Passaging was accomplished by treating confluent cultures with a 0.25-mg/ml solution of trypsin in EBSS containing 0.2 mg/ml of EDTA for 4 min at between 4 and 15°C to detach cells. The proteolytic activity was quenched by the addition of fetal calf serum, and the resulting cell suspension was centrifuged; the cell pellet was resuspended in fresh growth medium, and 1-ml aliquots were dispensed into culture flasks containing 9 ml growth medium.

2.3. Hormone Additions to T-470 Cells

Approximately 10^6 cells were transferred in 1.0-ml volumes to 25-cm² flasks containing 3.0 ml of growth medium. Hormone treatment of cells was initiated 48 hr later, after the cells were about 75% confluent, by replacement of growth medium with medium consisting of MEM, fetal calf serum (to 10% by volume) that had been treated with dextran-coated charcoal (Horwitz et al., 1975) to remove endogenous steroids, 0.1 mM MEM nonessential amino acids, 0.2 U/ml of Lente insulin, 10^{-8} M T_3, and in most experiments, 1 mM sodium pyruvate. Steroids were added to the medium as concentrated solutions in absolute ethanol. The total ethanol concentration in the medium did not exceed 0.2%, and the same amount of ethanol was added to all of the media used in a given experiment. Prolactin was added in alkaline solution. Experimental media were replaced every 48 hr.

2.4. Pulse-Labeling of Cultures with [2-¹⁴C]Acetate

After the appropriate period of hormone treatment, the experimental media were removed and the cultures were washed with EBSS. Two or three ml of MEM containing 1.0 μCi/ml [¹⁴C] acetate, 0.1 mM MEM nonessential amino acids, 1 mM pyruvate, and 0.2 U/ml of insulin were added to each flask without serum and incubated for 1 hr at 37°C. An aliquot (20 μl) of medium was counted before addition to the flasks. Similarly, [³H]palmitate was dissolved in the same medium (0.5 μCi/ml) for pulse-labeling of cells. Additions to each flask were timed so that incubations could be stopped at precisely 60 min. Media were removed by aspiration into a collection flask, the cultures were washed with EBSS, and detached as described above, and centrifuged.

2.5. Analysis of Cultured Cells for [¹⁴C and ³H]Lipids, Triglyceride Content, and DNA

Cell pellets were extracted with chloroform-methanol (2 : 1) (Radin, 1969) for lipids. An aliquot of the washed organic phase was counted for radioactivity and the remaining extract was applied to silica-gel thin-layer chromatography plates for separation of several lipid fractions. The insoluble material from the chloroform–methanol extraction was analyzed for DNA by the method of Burton (1956).

Thin-layer chromatograms were developed with petroleum ether–diethyl ether–glacial acetic acid (80 : 20 : 1). A standard lipid mixture containing lecithin, cholesterol, palmitic acid, triolein, and cholesteryl palmitate was chromatographed on the same plate with the samples. Each of the reference compounds was separated from each other in the order listed, moving from the point of application. After development of the chromatograms and evaporation of the adhering solvents, the plates were sprayed with a 0.1 mg/ml aq. solution of rhodamine 6G. Lipid fractions were eluted according to the location of the standards observed under UV light. The silica gel containing triglycerides was transferred to a Pasteur pipette that had been plugged with glass wool. Triglycerides were eluted with diethyl ether. Aliquots were counted for radioactivity, and the remaining material was assayed colorimetrically for its glycerol content after saponification (Galletti, 1967). Other lipid fractions from the chromatogram were counted for radioactivity directly by transferring the silica gel containing the material directly into a counting vial; the scintillation fluid served to elute the material as well as to provide the medium for counting the radioactivity. In some experiments all zones from the chromatogram were eluted directly in the scintillation vials.

In some cases the saponified triglycerides were anlyzed for their fatty acid composition by gas chromatography. Fatty acids were extracted from the saponification mixture with ethyl acetate after acidification of the mixture. The solution in ethyl acetate was washed with water and the organic phase was dried over granular sodium sulfate. After evaporation of the ethyl acetate, the residue was dissolved in a small volume of chloroform for injection into the gas chromatograph. The gas chromatograph was standardized by recording the retention times of several fatty acids ranging from C_8 to C_{22}. Separation was achieved isothermally at 200°C on a column containing 10% SP-216-PS on 100/120 mesh Supelcoport (Supelco, Inc., Bellefonte, PA). Fatty acids were detected by a flame ionization detector. Quantification of recorded peaks was by weight.

2.6. Electron Microscopy

Methods used for preparation and evaluation of the cultured cells for electron microscopy were those described previously (Verhage *et al.*, 1980). The sections were cut with a diamond knife, stained with lead citrate and uranyl acetate, and examined on a Zeiss EM9A microscope.

3. Results

3.1. Ultrastructural Appearance of Lipid in T-47D cells

A survey of thick (1 μm) sections of T-47D cells cultured in the basal medium containing PRL, T_3, cortisol (F), and insulin (I) revealed that cells

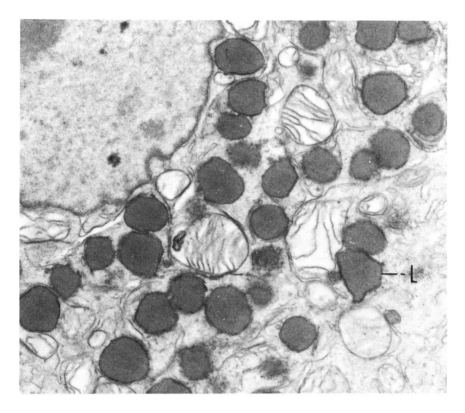

Figure 1. Electron micrograph of a T-47D cell incubated for 8 days in Eagle's MEM containing 10% by volume of charcoal-stripped fetal calf serum, 0.1 mM MEM nonessential amino acids, 1.0 mM pyruvate, 0.2 U/ml Lente insulin, 10^{-8} M triiodothyronine, 1.0 μg/ml ovine prolactin (NIH-P-S10), 1.0 μg/ml of progesterone, 1.0 μg of cortisol, and 1.0 μl/ml of ethanol. Magnification, 14,000. Note the darkly stained lipid droplets (L) that are about the same size as the mitochondria.

grown to confluency contained lipid droplets. Those grown in medium that also contained progesterone (P) had a greater abundance of similar droplets. The appearance of these droplets by electron microscopy is shown in Fig. 1.

3.2. Characteristics of the Response to Progesterone

The response to P, determined in the presence of I, PRL, T_3, and F, was measured as the ability of cells in confluent cultures to incorporate ^{14}C from [2-^{14}C]acetate into several lipid fractions. The total incorporation of ^{14}C into the lipid extract was not increased by P. However there was an increase in the incorporation of ^{14}C into the triglyceride fraction (Table I). A decrease occurred in the fraction containing phospholipids, i.e., the most polar fraction by thin-layer chromatography.

Table I. Triglyceride Synthesis in Confluent T-47D Cell Cultures as a Function of Treatment and Time

| Treatment | Percent of [^{14}C]lipid as triglycerides[a] after hormone treatment (hr) | | | |
	12	24	48	72
Control	21.0 ± 1.9	21.0 ± 1.3	22.0 ± 1.0	17.0 ± 0.5
Lactogenic	25.0 ± 1.4	19.6 ± 1.4	24.0 ± 2.2	22.0 ± 1.5
Progesterone + lactogenic	24.0 ± 0.9	28.0 ± 0.3	35.0 ± 1.0	42.0 ± 1.3

[a] Proportion of ^{14}C-labeled triglycerides calculated by dividing the cpm in the triglyceride fractions by the total cpm recovered from the chromatograms after TLC. Values represent the mean of triplicate determinations ± 1 standard error.

Accumulation of triglycerides (TG) measured colorimetrically after separation of lipids by thin-layer chromatography was also increased by the presence of progesterone (Fig. 2A), and this was not due to an increase in the number of cells per culture since the amount of DNA in these cultures was not changed by the presence of P (Fig. 2B). By comparison of Table I and Fig. 2A, it is apparent that while the time-course of ^{14}C accumulation from [2-^{14}C]acetate in TG is relatively uniform, the accumulation of TG occurred most rapidly during the last 24 hr in culture.

To determine whether the hormones PRL or F are required for the TG response to progesterone, medium containing only the hormones insulin and T$_3$ was tested with each of the other hormones added separately or in combination. Triglycerides were measured after 72-hr incubation with hormones. The data in Fig. 3A show that addition of P alone provided the same stimulus that was observed with combinations of P with F and PRL. Further attempts to demonstrate an effect of cortisol in doses of 10^{-11}–10^{-5} M were unsuccessful. Incorporation of ^{14}C from [2-^{14}C]acetate into total extractable lipids was not changed (Fig. 3B), nor was the DNA per culture at the time the cells were harvested (Fig. 3C).

3.3. Reversibility of the Progesterone Effect

In this experiment the rate of ^{14}C incorporation from [^{14}C]acetate into TG was compared in cultures grown in medium containing I, T$_3$, and P (1 µg/ml) with control cultures grown without P. After 48 hr, cultures grown with P had 212–225% the amount of TG in controls. Cultures with P that were grown for an additional 96 hr in the presence of P had 428% of the TG in control cultures at the same time period. On the other hand, cultures in which P was withdrawn after 48 hr had only 155% as much TG as control cultures 96 hr later. This indicates reversibility of the progesterone effect in these cells. In this experiment also the ^{14}C in total extractable lipids was not different in cells exposed to P than in those that were not.

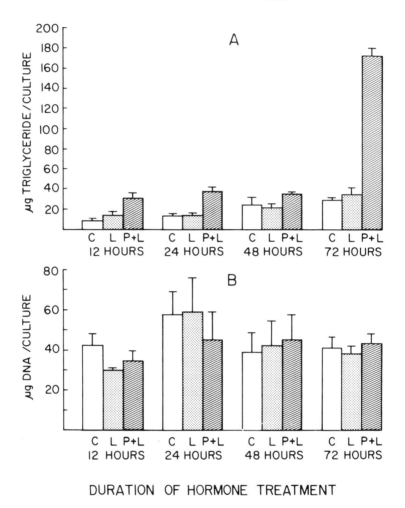

Figure 2.   Triglyceride (A) and DNA (B) content of confluent T-47D cell cultures after 12–72 hr of hormone treatment. Values represent the mean of triplicate determinations ± SEM.

3.4. Steroid Structural Requirements for Triglyceride Stimulation

A number of steroids were tested for their ability to promote the accumulation of triglycerides over a 72-hr period of exposure. All steroids were tested at a concentration of 3.2 μM (equivalent to 1.0 μg/ml P). The results are shown in Table II as a percentage of that obtained in the same medium without progesterone. The value shown for each steroid is the average of a minimum of three replicate assays. Although cortisol, as previously described, had no promotional activity,

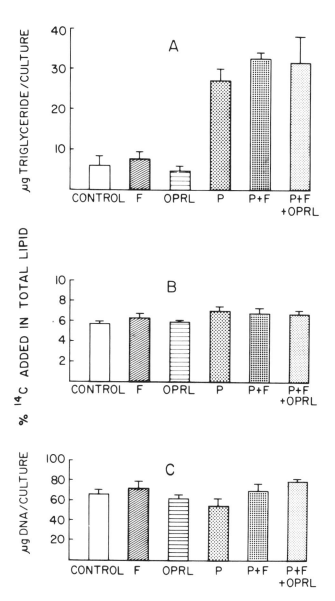

Figure 3.   Triglyceride levels (A), rates of total [^{14}C]acetate (B), and DNA levels (C) in confluent T-47D cell cultures after treatment with various hormones, all at concentrations of 1.0 μg/ml, for 72 hr. F, cortisol; OPRL, ovine prolactin; P, progesterone. Values represent the mean of triplicate determinations ± SEM.

Table II. Steroid Structural Requirements for
Stimulation of Triglyceride Accumulation in
T-47D Cells

Steroid	Percent of control ($\pm$ SE)
Deoxycorticosterone	107 $\pm$ 5
17-hydroxyprogesterone	110 $\pm$ 17
11α-hydroxyprogesterone	129 $\pm$ 26
Cortisol acetate	136 $\pm$ 26
Dexamethasone	166 $\pm$ 6
Dihydrotestosterone	171 $\pm$ 22
Aldosterone	185 $\pm$ 3
20α-hydroxy-4-pregnen-3-one	202 $\pm$ 27
20β-hydroxy-4-pregnen-3-one	204 $\pm$ 8
R5020	227 $\pm$ 8
Norethindrone	284 $\pm$ 28
Corticosterone acetate	288 $\pm$ 56
Progesterone	315 $\pm$ 32
Cyproterone acetate	385 $\pm$ 58

dexamethasone had a significant effect. Among glucocorticoids, corticosterone acetate was notable in its activity, being similar to that of P. The 20α-reduction products of P were about half as active as P, and synthetic progestogens were in the same range as P.

3.5. Triglyceride Secretion vs. Accumulation within T-47D Cells

To determine whether TG synthesized during a 72-hr incubation with and without P was secreted by the cultured cells, the incubation was continued for another 11 hr after the 1-hr pulse with [^{14}C]acetate. Medium and cells were assayed separately for ^{14}C in total lipids and in the TG fraction. The cells contained between 4.0 and 6.5% of the ^{14}C added in extractable lipids, but the medium contained less than 0.3% of the ^{14}C dose. Triglyceride levels were too low in the medium to be detected, but they accounted for about 40% of total lipid in cells exposed to P.

3.6. Dose Response to Progesterone

Confluent cultures of T-47D cells were exposed to media containing I, T_3, and PRL, and different dosage of P. Some cultures were also treated with a low concentration of F (10^{-9} M). The response to P was significantly increased at 10^{-8} M, and reached a maximum at 10^{-5} (Fig. 4). Cell death occurred at 10^{-4}

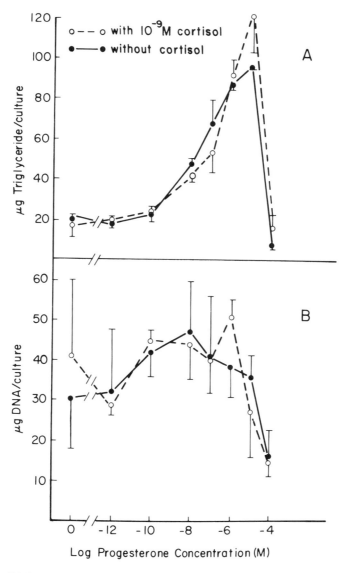

Figure 4. Triglyceride (A) and DNA (B) levels in confluent cultures of T-47D cells after treatment with various dosages of progesterone in the presence of ovine prolactin (1.0 μg/ml) and with or without 10^{-9} M cortisol. Values represent the mean of triplicate determinations ± SEM.

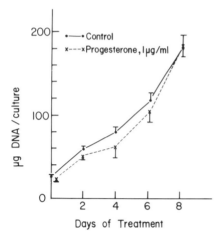

Figure 5. Effect of progesterone on the growth rate of T-47D cells. Replicate cultures in log phase were grown in the basal medium containing insulin (0.2 U/ml) and triiodothyronine (10^{-8} M) with or without progesterone (1.0 μg/ml). Values represent the mean of five determinations ± SEM.

M. Cortisol at 10^{-9} M had no effect on accumulation of TG by T-47D cells at any concentration of P.

3.7. Effect of Progesterone on Cell Proliferation during Log-Phase Growth

Cultures were grown in medium containing I and T_3 with or without 1.0 μg/ml P. Progesterone had no effect on the rate of growth of the cells (Fig. 5).

3.8. Cell-Specificity of the Response to Progesterone

Mammary cell lines MCF-7 and HBL-100 as well as fibroblast cell lines 28 and 857 were tested for their ability to respond to P (1.0 μg/ml) with an increase in accumulation of TG during a 72-hr incubation period after confluency. Under these conditions none of the above cell lines accumulated an increase in TG in the presence of P.

3.9. Effect of Pyruvate on Incorporation of [^{14}C]Acetate into Triglycerides

Although the incorporation of ^{14}C from [2-^{14}C]acetate into total extractable lipids was not changed, ^{14}C in TG as a percentage of that in total lipids was decreased by 1.0 mM pyruvate, when P was present in medium. Without P, no effect was noted (Table III).

Table III. Effect of Pyruvate on Incorporation of [^{14}C]Acetate
into Triglycerides

Treatment		^{14}C incorporation into total lipids, dpm $\times$ 10^{-5}	Percent as triglycerides
Pyruvate, 1.0 μM	Progesterone, 3.2 μM		
No	No	4.0 ± 0.6	13.4 ± 1.2
Yes	No	4.8 ± 0.2	12.3 ± 0.4
Yes	Yes	4.0 ± 0.2	23.7 ± 0.8
No	Yes	5.0 ± 0.1	35.2 ± 1.3

3.10. Accumulation of Medium-Chain Fatty Acids in Triglycerides of T-47D Cells

A comparison was made between the fatty acid pattern in TG extracted from T-47D cells with that in TG extracted from normal human milk. Some of the major fatty acids found were quantified. The proportional amounts found are presented in Table IV. In cells cultured in the presence of only the hormones I and T_3 with or without 0.1% ethanol there is an abundance of lauric acid (12 : 0) and little myristic (14 : 0). Triglycerides extracted from cells that had been incubated with P or cyproterone acetate contained much less lauric and more normal amounts of palmitic (16 : 0) and oleic (18 : 1) acids.

3.11. Incorporation of [^{3}H]Palmitate into Lipids in T-47D Cells

Lipids were extracted from the washed cells immediately after termination of exposure to 0.5 μCi of [^{3}H]palmitate (72 hr after incubation with hormones).

Table IV. Proportional Amounts of Several Fatty Acids in Triglycerides from
T-47D Cells

	Proportion fatty acid			
	12 : 0	14 : 0	16 : 0	18 : 1
Human milk (Smith and Abraham, 1975)	9.8	11.2	29.2	49.8
Human milk (this study)	6.9	4.8	17.6	70.6
Serum	0.7	6.4	40.9	52.0
T-47D Cells				
Control	31.7	1.7	39.2	27.3
Ethanol (0.1%)	31.4	2.2	39.7	26.7
Progesterone (3.2 μM)	2.7	5.4	48.2	43.7
Cyproterone acetate (3.2 μM)	3.4	5.2	39.8	51.5

Table V. [³H]Palmitate Incorporation into Triglycerides by T-47D Mammary Tumor Cells

Pulse time	Medium	Total[³H]-lipid/flask dpm × 10⁻⁵	Percent as triglycerides	Percent as free fatty acids	Percent as phospholipids
1 hr	Progesterone	4.0 ± 0.4	37.2 ± 0.8	4.0 ± 0.6	52.4 ± 2.4
	Control	3.0 ± 0.4	15.2 ± 2.1	7.6 ± 1.1	70.1 ± 10.0
2 hr	Progesterone	6.4 ± 0.6	37.0 ± 1.6	7.0 ± 0.8	50.3 ± 5.0
	Control	6.2 ± 0.2	12.8 ± 0.8	9.0 ± 0.8	67.3 ± 2.7
4 hr	Progesterone	8.7 ± 0.4	47.8 ± 1.6	5.6 ± 0.7	36.6 ± 7.9
	Control	8.9 ± 0.5	7.4 ± 1.4	11.2 ± 0.9	74.9 ± 7.6

Although the incorporation in total lipids increased with duration of exposure to the labeled precursor, P did not increase the rate of incorporation except possibly initially (Table V). However, P did cause greater accumulation of [³H]palmitate in TG, and the difference between cells with and without P exposure became greater as the pulse time with [³H]palmitate increased. In cells not exposed to P, [³H]palmitate in TG decreased twofold within 4 hr. Cells exposed to P had an increase in [³H]palmitate in TG that roughly corresponded to the decrease in the phospholipid fraction.

4. Discussion

The relation between the findings with regard to P stimulation of triglyceride accumulation in T-47D cells and the normal physiology of the human mammary gland is not clear, particularly because these cells do not respond to glucocorticoids or prolactin. Cameron *et al.* (1974) found that the normal rat mammary gland responded to the combination of I, PRL, and F with maximal lipid biosynthesis *in vitro.* Apparently, despite the presence of both glucocorticoid (Horwitz *et al.,* 1978) and PRL (Shiu, 1979) receptors, the stimulatory pathways in T-47D cells are incomplete with respect to these hormones.

An interesting question is whether the response to progesterone represents a normal response of human mammary epithelium to this hormone, or a process peculiar to this particular tumor cell. The lack of triglyceride accumulation in other cell types under conditions promoting the process in T-47D cells would seem to obviate the possibility that this is a nonspecific process related to an aging or toxic effect on cultured cells. Indeed, the maximal effect was achieved at a concentration of P that caused no decrease in cell numbers in confluent cultures, and log-phase growth was not affected by a highly stimulatory dose. The formation of triglycerides containing medium-chain-length fatty acids also shows that lipid synthesis in the T-47D cells is related to the function of the mammary epithelium from which they were derived. Inhibition of [^{14}C]acetate incorporation by pyruvate is also characteristic of mammary epithelium (Hirsch *et al.*, 1954).

Although P is generally considered an inhibitor of lactation, in fact when P is secreted (Chatterton *et al.*, 1975) or given to animals after milk secretion has been fully initiated (Walker and Matthews, 1949), it does not decrease the production of milk. Women who have been given progestogens for fertility control during lactation do not usually experience a decrease in milk production as long as lactation has become well-established and estrogens are not included, and they may actually have increased milk production (Hull, 1981). Does this represent a partial agonistic activity of P interacting with the glucocorticoid receptor to which it binds with relatively high affinity (Lindenbaum and Chatterton, 1981)? The answer to this question, based on response data, is that it does not. Progesterone was more active in stimulating accumulation of triglycerides than was cortisol or dexamethasone. Likewise, it would not seem to have partial agonistic action through the mineralocorticoid receptor since aldosterone and deoxycorticosterone were both clearly less active than P. Studies have yet to be completed relating binding affinity to the triglyceride stimulatory activity of the structurally related steroids, but it would seem that P is acting through a receptor that is relatively specific for progestins. Such a receptor has been identified in these cells (Horwitz *et al.*, 1978). In this respect the response being measured *in vitro* is one of the few responses to steroid hormones that can be studied entirely *in vitro*. Westley and Rochefort (1980) have observed the production of a secretory protein in the MCF-7 cell line in response to estradiol-17β. The steroid-induced responses require much more time for formation of a measurable product than, for example, the fatty acid synthesis induced in the MCF-7 cells by insulin (Monaco and Lippman, 1977).

The incorporation of ^{14}C from [^{14}C]acetate in TG per flask increased during the 1-hr pulse throughout the 72-hr course of the incubation. The incubation began 48 hr after seeding the flasks and growth occurred throughout the incubation period such that the incorporation per μg DNA remained constant. Nevertheless, the accumulation of triglycerides increased markedly only within the last 24 hr of the 72-hr incubation when P was present in the medium. The reason for greater accumulation than apparent formation from [2-^{14}C]acetate may be in the utilization of preformed fatty acids either from hydrolysis of triglycerides in the medium or from hydrolysis of cellular phospholipids. The decrease in proportion of short-chain fatty acids after stimulation of triglyceride by progesterone supports this supposition. In incubations containing [^{3}H]palmitate, P promoted the retention of [^{3}H]palmitate in triglycerides while [^{3}H]palmitate that had been incorporated into phospholipids decreased with longer incubations after addition of [^{3}H]palmitate. While no conclusions can be drawn about the mechanisms involved from this experiment, the effect of P could be explained by a decrease in triacylglycerol lipase activity (i.e., an insulin-like action on this enzyme) or by an increase in phospholipase activity.

In any case, the action of progesterone differs from that of insulin in that the synthesis of total fatty acids in all lipid fractions was not increased. The effect appears to be only on the preferential accumulation of triglycerides, a process that could be controlled through hormone-sensitive lipase or phospholipase.

ACKNOWLEDGMENTS. The studies reported in this paper were supported by a grant from the NIH, HD-13975. The authors also thank Sara Day Cheesman for preparation of the manuscript.

REFERENCES

Bailey, J. M., Howard, B. V., and Tillman, S. F., 1973, Lipid metabolism in cultured cells, *J. Biol. Chem.* **248**:1240.

Barnawell, E. B., 1967, Analysis of the direct action of prolactin and steroids on mammary tissue of the dog in organ culture, *Endocrinology* **80**:1083.

Burton, K. A., 1956, A study of the conditions and mechanism of the diphenylamine reaction for the colorimetric estimation of deoxyribonucleic acid, *Biochem. J.* **62**:315.

Cameron, J. A., Rivera, E. M., and Emergy R. M., 1975, Hormone-stimulated lipid synthesis in mammary culture, *Am. Zool.* **15**:285.

Chatterton, R. T., Jr., King, W. J., Ward, D. A., and Chien, J. L., 1975, Differential responses of prelactating and lactating mammary gland to similar tissue concentrations of progesterone, *Endocrinology* **96**:861.

Chatterton, R. T., Jr., Harris, J. A., King, W. J., and Wynn, R. M., 1979, Ultrastructural alterations in mammary glands of pregnant rats after ovariectomy and hysterectomy: Effect of adrenal steroids and prolactin, *Am. J. Obstet. Gynecol.* **133**:694.

Dils, R., Forsyth, I., and Strong, C. R., 1972, Hormonal control of the synthesis of milk fatty acids by rabbit mammary gland, *J. Physiol.* **222**:94P.

Galletti, F., 1967, An improved colorimetric micromethod for determination of serum glycerides, *Clin. Chim. Acta* **15**:184.

Greenbaum, A. L., and Darby, F. J., 1964, The effect of adrenalectomy on the metabolism of the mammary glands of lactating rats, *Biochem. J.* **91**:307.

Halestrap, A. P., and Denton, R. M., 1973, Insulin and the regulation of adipose tissue acetyl-Coenzyme A carboxylase, *Biochem. J.* **132**:509.

Hallowes, R. C., Wang, D. Y., Lewis, D. J., Strong, C. R., and Dils, R., 1973, The stimulation by prolactin and growth hormone of fatty acid synthesis in explants from rat mammary glands, *J. Endocrinol.* **57**:265.

Hirsch, P. F., Baruch, H., and Chaikoff, I. L., 1954, The relation of glucose oxidation to lipogenesis in mammary tissue, *J. Biol. Chem.* **210**:785.

Horwitz, K. B., and McGuire, W. L., 1975, Specific progesterone receptors in human breast cancer, *Steroids* **25**:497.

Horwitz, K. B., Costlow, M. E., and McGuire, W. L., 1975, MCF-7: A human breast cancer cell line with estrogen, androgen, progesterone and glucocorticoid receptors, *Steroids* **26**:785.

Horwitz, K. B., Zava, D. T., Thilaghar, A. K., Jensen, E. M., and McGuire, W. L., 1978, Steroid receptor analyses of nine human breast cancer cell lines, *Cancer Res.* **38**:2434.

Hull, V. J., 1981, The effects of hormonal contraceptives on lactation: Current findings, methodological considerations, and future priorities, *Stud. Fam. Plann.* **12**:134.

Jacobs, R. A., Sly, W. J., and Majerus, P. W., 1973, The regulation of fatty acid biosynthesis in human skin fibroblasts, *J. Biol. Chem.* **248**:1268.

Katz, J., Wals, P. A., and Van de Velde, R. L., 1974, Lipogenesis by acini from mammary gland of lactating rats, *J. Biol. Chem.* **249**:7348.

Keydar, I., Chen, L., Karby, S., Weiss, F. R., Delarea, J., Radu, M., Caitcik, M., and Brenner, H. J., 1979, Establishment and characterization of a cell line of human breast carcinoma origin, *Eur. J. Cancer* **15**:659.

Kyriakou, S. Y., and Kuhn, N. J., 1973, Lactogenesis in the diabetic rat, *J. Endocrinol.* **59**:199.

Lawrence, R. A., 1980, Maternal diabetes, in: *Breast-Feeding. A Guide for the Medical Profession* (R. A. Lawrence, ed.), pp. 240–242, C. V. Mosby Co., St. Louis.

Leader, D. P., and Barry, J. M., 1969, Increase in activity of glucose-6-phosphate dehydrogenase in mouse mammary tissue cultured with insulin, *Biochem. J.* **113**:175.

Lindenbaum, M., and Chatterton, R. T., Jr., 1981, Interaction of steroids with dexamethasone-binding receptor and corticosteroid-binding globulin (CBG) in mammary gland of the mouse in relation to lactation, *Endocrinology* **109**:363.

Martin, R. J., and Baldwin, R. L., 1971, Effects of alloxan diabetes on lactational performance and mammary tissue metabolism in the rat, *Endocrinology* **88**:863.

McLean, P., Greenbaum, A. L., and Gumaa, K. A., 1972, Constant and specific proportion groups of enzymes in rat mammary gland and adipose tissue in relation to lipogenesis, *FEBS Lett.* **20**:277.

Monaco, M. E., and Lippman, M. E., 1977, Insulin stimulation of fatty acid synthesis in human breast cancer in long term tissue culture, *Endocrinology* **101**:1238.

Moretti, R. L., and Abraham, S., 1966, Effects of insulin on glucose metabolism by explants of mouse mammary gland maintained in organ culture, *Biochem. Biophys. Acta* **124**:280.

Nikolaeva, M. Y., and Balmukhanov, B. S., 1980, Effects of insulin on sections of rat's mammary gland, *Biofizika* **24**:860.

Norgren, A, 1967, Mammary development in alloxan diabetic rabbits injected with oestrone and progesterone, *Acta Univ. Lund. Section II,* No. 37.

Oka, T., Perry, J. W., and Topper, Y. J., 1974, Changes in insulin responsiveness during development of mammary epithelium, *J. Cell Biol.* **62**:550.

Olefsky, J. M., 1975, Effect of dexamethasone on insulin binding, glucose transport, and glucose oxidation of isolated rat adipocytes, *J. Clin. Invest.* **56**:1499.

Radin, N. S., 1969, Preparation of lipid extracts, in: *Methods of Enzymology,* Vol. 24 (J. M. Lowenstein, ed.), pp. 245–254, Academic Press, New York.

Raff, R. A., 1970, Induction of fatty acid synthesis in cultured mammalian cells: Effects of cyclo-heximide and X-rays, *J. Cell. Physiol.* **75**:341.

Robinson, A. M., and Williamson, D. H., 1977, Comparison of glucose metabolism in the lactating mammary gland of the rat *in vivo* and *in vitro.* Effects of starvation, prolactin, or insulin deficiency, *Biochem. J.* **164**:153.

Shiu, R. P. C., 1979, Prolactin receptors in human cancer cells in long-term tissue culture, *Cancer Res.* **39**:4381.

Smith, S., and Abraham, S., 1975, The composition and biosynthesis of milk fat, in: *Advances in Lipid Research,* Vol. 13 (R. Paoletti and D. Kritchevsky, eds.) pp. 195–239, Academic Press, New York.

Speake, B. K., Dils, R., and Mayer, R. J., 1975, Regulation of enzyme turnover during tissue differentiation: Studies on the effects of hormones on the turnover of fatty acid synthetase in rabbit mammary gland in organ culture, *Biochem. J.* **148**:309.

Sutter-Dub, M. T., Leclercq, R., Sutter, B. C. J., and Jacquot, R., 1974, Plasma glucose, pro-
    gesterone and immunoreactive insulin in the lactating rat, *Horm. Metab. Res.* **6:**297.
Verhage, H. G., Bareither, M. L., Jaffe, R. C., and Akbar, M., 1980, Cyclic changes in ciliation,
    secretion, and cell height of the oviductal epithelium in women, *Am. J. Anat.* **156:**505.
Walker, S. M., and Matthews, J. I., 1949, Observations on the effects of prepartal and postpartal
    estrogen and progesterone treatment on lactation in the rat, *Endocrinology* **44:**8.
Westley, B., and Rochefort, H., 1980, A secreted glycoprotein induced by estrogen in human breast
    cancer cell lines, *Cell* **20:**353.
Williamson, D. H., McKeown, S. R., and Elic, V., 1974, Metabolic interactions of glucose,
    acetoacetate, and insulin in mammary gland slices of lactating rats, *Biochem. J.* **150:**145.

5

Prolactin and Casein Gene Expression in the Mammary Cell

*Louis-Marie Houdebine, Jean Djiane,
Bertrand Teyssot, Jean-Luc Servely,
Paul A. Kelly, Claude Delouis,
Michèle Ollivier-Bousquet, and Eve Devinoy*

1. Introduction

The onset of milk synthesis and secretion is the result of complex and multiple processes which operate during pregnancy and at parturition. Before pregnancy, the mammary gland is restricted to a few duct cells. During pregnancy, under the influence of estrogens, progesterone, and growth factors, the secretory cells progressively appear. They are organized in an epithelium forming a large number of alveoli. At the end of pregnancy, many alveolar cells are present and the development of the mammary gland is more or less complete according to species. After parturition, when milk secretion is triggered, the alveolar cells become polarized and hypertrophic. This transformation corresponds to the activation of the cells which have to elaborate and secrete huge amounts of proteins, lipids, and carbohydrates throughout lactation. Thus before being fully active, the mammary gland has been subjected to at least three types of transformation: (1) a cell multiplication which leads to the formation of alveoli, (2) an activation of specific

Louis-Marie Houdebine, Jean Djiane, Bertrand Teyssot, Jean-Luc Servely, Claude Delouis, Michèle Ollivier-Bousquet, and Eve Devinoy • Laboratoire de Physiologie de la Lactation, Institut National de la Recherche Agronomique, C.N.R.Z., 78350 Jouy-en-Josas, France *Paul A. Kelly* • Département d'Endocrinologie Moléculaire, Centre Hospitalier de l' Université Laval, Quebec, Canada.

genes directly involved in the elaboration of milk, and (3) an organization of the alveolar cells which become enriched in cellular organelles involved in the bulky production and secretion of milk. After weaning, the alveolar cells disappear until the next pregnancy.

All these processes are under the control of several hormones in which prolactin plays the major stimulating role. Its actions are modulated by ovarian steroids which favor mammary gland development while preventing the induction of milk synthesis. The action of prolactin is also greatly amplified by glucocorticoids which are by themselves not inducers (Denamur, 1971).

Milk contains large amounts of lipids, carbohydrates, and proteins, most of them being elaborated in the alveolar cells, using precursors captured from the circulating blood. The major milk proteins in most species are caseins which constitute a family of proteins. These proteins have been extensively studied in ruminants and the primary sequence of αs_1-, αs_2-, β-, and κ-caseins has been determined in cow milk (Mercier et al., 1972). These proteins contain about 200 aminoacids, with a signal peptide which is absent in the secreted proteins (Mercier and Gaye, 1980). Apart from caseins, milk contains minor proteins including α-lactalbumin and β-lactoglobulin, also present in variable amounts according to species.

The mammary gland is therefore an excellent biological system for the study of the hormonal control of the expression of specific genes. In this respect, the major particularity of the mammary cell rests on the fact that the expression of specific genes is under the control of a protein hormone, the steroids playing only the role of modulators. The present report summarizes the experimental data obtained in the rabbit. This includes the study of prolactin receptors, the determination of casein synthesis and casein mRNA concentration, the measurements of casein gene transcription in isolated nuclei, and the identification of a putative intracellular relay playing the role of a second messenger carrying the hormonal information from prolactin receptors on the plasma membrane to casein genes. A summary of our previous work has already been published (Houdebine, 1980b,c; Houdebine et al., 1982). This kind of study has also been carried out by other groups working in mouse (Mehta et al., 1980), rat (Qasba et al., 1981; Rosen et al., 1981), and guinea pig (Burditt et al., 1981).

2. The Expression of Casein Genes during Pregnancy and Lactation

In the rabbit, the development of the mammary gland starts around day 8 of pregnancy and is almost completed at parturition. Until day 18 of pregnancy, no casein synthesis is detectable in the mammary gland. After this critical period, milk becomes visible in the tissue and casein synthesis is progressively increased while the development proceeds. One week before parturition the relative casein synthesis is already at a high level, but the alveolar cells are not yet fully polarized and hypertrophic (the Golgi apparatus is only partly developed and the RNA/DNA ratio remains approximately 1 compared with 3–4 when lactation is fully established). Casein mRNA concentration is increased after day 18 while casein synthesis is induced. However, a significant amount of casein mRNA is already present in early pregnancy and even in the virgin animal. This suggests that the accumulation of casein mRNA during the second half of pregnancy may be responsible for the induction of casein synthesis during this period but that, in early pregnancy and before pregnancy, the mammary cell is not capable of translating this casein mRNA efficiently. It may be concluded that the activation of casein gene expression during the natural pregnancy–lactation cycle is not an all-or-none phenomenon but that the gland is very progressively prepared to synthesize milk (Shuster *et al.*, 1976). Roughly similar conclusions were drawn from studies carried out in the rat (Rosen *et al.*, 1975; Nakhasi and Qasba, 1979) and the guinea pig (Burditt *et al.*, 1981). In the rabbit, a relatively precocious milk synthesis takes place owing to the early initial drop of progesterone during the second half of pregnancy (Houdebine *et al.*, 1982) and to the exceptional sensitivity of this species towards prolactin.

3. Relation between Casein Synthesis and Casein mRNA Concentration

3.1. The Actions of Prolactin, Glucocorticoids, and Progesterone in the Pseudopregnant Rabbit

In the rabbit, pseudopregnancy can be easily induced by the mating of a female with a vasectomized male or by injecting a mature female with hCG. During the 15 days following the induction of pseudopregnancy, the growth of the mammary gland is similar to that during normal pregnancy. After this period of pseudopregnancy, the corpus luteum involutes, the progesterone level drops, and the mammary gland regresses. The development and the synthetic activity of the mammary tissue become dependent upon hormones injected into the animal or added to the culture medium of explants or isolated cells. It is worth noting that the rabbit is the only species so far studied in which injections of prolactin alone can elicit a significant growth and activity of the mammary gland.

Injections of prolactin into pseudopregnant or pregnant rabbits 15 days after mating induce casein synthesis, as a function of the dose injected (Houdebine and Gaye, 1975; Houdebine, 1976). When moderate doses of prolactin are injected, casein synthesis may be amplified or inhibited by the simultaneous injections of glucocorticoids or progesterone, respectively (Devinoy and Houdebine, 1977; Devinoy *et al.*, 1979; Houdebine and Gaye, 1975; Houdebine, 1976). Glucocorticoids in the absence of prolactin (a situation provoked by injecting a

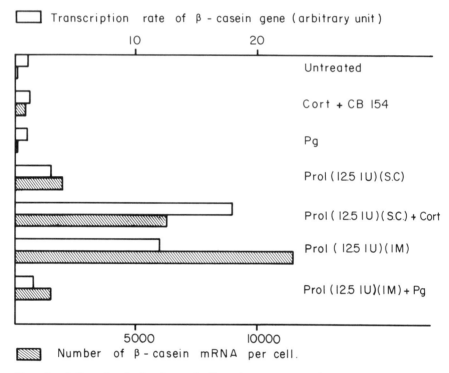

Figure 1.   Actions of prolactin, glucocorticoids, and progesterone on the accumulation of β-casein mRNA and the transcription of β-casein gene. The β-casein mRNA sequences in total mammary RNA and in mercurated RNA neosynthesized in isolated nuclei were evaluated with a β-casein DNA probe. Hormones were injected twice daily into pseudopregnant rabbit for 4 days. Prolactin (Prol) was injected subcutaneously (S.C.) or intramuscularly (I.M.). Hydrocortisone acetate was injected subcutaneously (7.5 mg/injection). Progesterone was injected intramuscularly (5 mg/injection). Bromocriptine (CB 154) was injected subcutaneously (2 mg/injection).

dopaminergic drug, bromocriptine CB 154), are essentially inactive. The determination of casein mRNA concentrations with a labeled cDNA probe revealed that in all cases the induction of casein synthesis is accompanied by a simultaneous accumulation of casein mRNA (Fig. 1). This strongly suggests that casein mRNA levels are one of the factors limiting the rate of casein synthesis.

3.2.   The Action of Prolactin in the Virgin Rabbit

Injections of prolactin into nulliparous mature animals do not trigger the development of the mammary gland, unless the female has been pretreated by estrogens or is subjected to a long and acute treatment by prolactin. Injections of prolactin in the virgin mature rabbit were unable to initiate casein synthesis, although they led to an accumulation of casein mRNA. This indicates that the duct cells contain casein mRNA and can understand the prolactin message but that, unlike the alveolar cells, they are unable to translate casein mRNA to a significant degree (Houdebine, 1977b). A similar observation has also been recorded in cultured mouse explants (Vondehaar *et al.*, 1978).

3.3. The Action of Prolactin, Glucocorticoids, and Progesterone in Mammary Organ Cultures

Mammary tissue explanted during pregnancy or pseudopregnancy is fully responsive to lactogenic hormones when cultured in the presence of synthetic media. Addition of prolactin to 199 medium induces both the accumulation of casein mRNA and casein synthesis. Cortisol amplifies prolactin action but is totally inactive in the absence of the protein hormone (Devinoy *et al.*, 1978). These observations confirm the data obtained *in vivo* concerning the role and the action of both hormones. The same hormonal specificity has also been observed in isolated epithelial mammary cell in culture (Teyssot *et al.*, 1981b). An essentially similar conclusion was drawn from experiments carried out in the rat (Matusik and Rosen, 1978) and mouse (Mehta *et al.*, 1980; Takemoto *et al.*, 1980).

4. The Control of Mammary Cell Hypertrophy

The hypertrophy of the mammary epithelial cells is an essential event in the induction of milk synthesis owing to the fact that these cells must elaborate considerable quantities of macromolecules from their precursors. In all species, the RNA/DNA ratio (i.e., the number of ribosomes per cell) remains low until parturition, thus until the drop of progesterone and the surge of prolactin. Similarly, the development of the intracellular network of membranes remains limited during pregnancy. In the pseudopregnant rabbit, injections of prolactin induce not only the appearance of new alveoli and the activation of casein gene expression but also the enhancement of the RNA/DNA ratio (Assairi *et al.*, 1974a), the formation of the intracellular membranes, and the binding of the casein-synthesizing polysomes to the endoplasmic reticulum (Houdebine, 1977a). Progesterone injected with prolactin strongly inhibits these effects of prolactin (Assairi *et al.*, 1974; Devinoy *et al.*, 1979). Glucocorticoids in the rabbit are unable to enhance the RNA/DNA ratio and to support the formation of the membrane–polysome complexes (Fig. 2) (Devinoy *et al.*, 1979; Fèvre and Houdebine, 1978). Progesterone thus appears to be a general inhibitor of prolactin actions, whereas glucocorticoids are amplifiers of only some of these actions. Throughout pregnancy in all species, progesterone prevents the induction of milk synthesis not only by limiting the release of prolactin by the hypophysis but also by activating other unknown

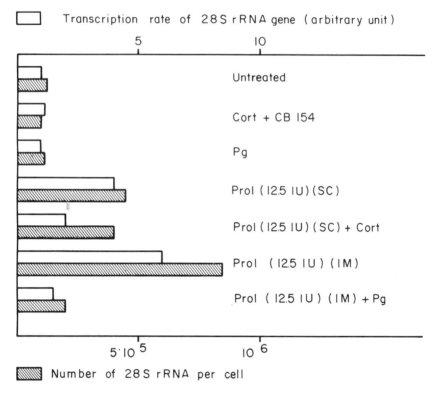

Figure 2. Actions of prolactin, glucocorticoids, and progesterone on the accumulation of 28 S ribosomal RNA and the transcription of the corresponding genes. The 28 S rRNA sequences in total mammary RNA and in mercurated RNA isolated from incubated nuclei were evaluated with a cDNA probe obtained by a reverse transcription of 28 S rRNA. Abbreviations as in Figure 1.

inhibitory mechanisms (Houdebine *et al.*, 1982). The steroid is also capable of easily counteracting the action of placental lactogenic hormones (in species in which they exist), which proved to be endowed with a rather limited lactogenic activity, at least in the ewe (Servely *et al.*, 1983).

It is worth mentioning that the cellular hypertrophy is at best marginally expressed in organ culture. Prolactin added to culture media induces a significant development of the Golgi apparatus but is unable to provoke the enhancement of the RNA/DNA ratio (Teyssot and Houdebine, 1981a). It is not known whether this discrepancy between the data obtained *in vivo* and in culture is due to deficiencies in the culture conditions or to the fact that the hypertrophy of the mammary cell requires multiple interactions of prolactin with various other target organs in the body of the animal.

5. The Hormonal Control of Casein mRNA Translation

Although the induction of casein synthesis is normally accompanied by an accumulation of casein mRNA, it is clear that a control of casein mRNA translation takes place in the mammary gland. This conclusion was drawn first from the fact that casein mRNA is present and can be accumulated in the virgin rabbit without being efficiently translated (Houdebine, 1977b). The study of thyroid

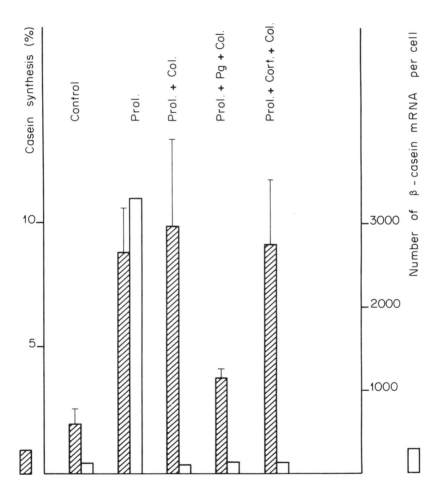

Figure 3.   Effects of prolactin, progesterone, glucocorticoids, and colchicine on the translation of casein mRNA. Prolactin (Prol) and Progesterone (Pg) were injected intramuscularly (12.5 IU and 5 mg/injection, respectively). Hydrocortisone acetate (Cort) and colchicine (Col) were injected subcutaneously). Casein synthesis is expressed as the percentage of proteins labelled in explants incubated at the end of the treatment with [^{14}C]amino acids and precipitated with anticasein antibody.

hormone action led to a similar conclusion: addition of thyroid hormone to culture media of explants increases casein synthesis in the presence of prolactin without supporting a further accumulation of casein mRNA (Houdebine *et al.*, 1978b).

In a study described in Section 9, it has been shown that colchicine and related drugs prevent the transmission of the prolactin message from receptor to genes, leading to an inhibition of casein synthesis and casein mRNA accumulation by prolactin in organ culture. Simultaneous injections of prolactin and colchicine into pseudopregnant rabbits result in a very significant induction of casein synthesis despite the inhibition of casein mRNA accumulation (Fig. 3) (Teyssot and Houdebine, 1981b). This clearly indicates that prolactin injected into pseudopregnant rabbits induces casein synthesis by stimulating simultaneously and independently the accumulation of casein mRNAs and their translation. Interestingly, progesterone inhibits strongly the translation of casein mRNA, whereas glucocorticoids are essentially ineffective (Fig. 3). This confirms that glucocorticoids amplify prolactin actions very selectively. This also points out the fact that throughout pregnancy casein mRNA accumulation and translation and the cellular hypertrophy are severely inhibited by progesterone. At parturition, the drop of progesterone and the resulting surge of prolactin allow the full expression of the mammary function.

The stimulation of casein mRNA translation by prolactin does not take place in organ culture, since both casein mRNA accumulation and casein synthesis are simultaneously inhibited by colchicine (see Section 9). This reinforces the idea that some of the prolactin actions are not, or are very poorly, expressed in culture.

The mechanism by which casein mRNA translation is controlled is not known. It might be related to the capacity of the mammary cell to trap amino acids, a phenomenon which was recently proved to be under the control of prolactin (Vina *et al.*, 1981).

6. Variations of Casein Gene Transcription and Casein mRNA Stability

The variations of casein mRNA accumulation suggest that the hormones control the transcription rate of casein genes. To evaluate this proposition, hormones were injected into pseudopregnant rabbits. Mammary nuclei were then isolated and incubated in the presence of mercurated cytidine triphosphate (HgCTP). The neosynthesized RNA thus mercurated was isolated by a SH-Sepharose column and the presence of β-casein mRNA sequences in the mercurated RNA eluted from the column by β-mercaptoethanol was quantified using a labeled cDNA probe obtained with purified β-casein mRNA.

Injections of prolactin into pseudopregnant rabbits provoke an acceleration of β-casein gene transcription in isolated mammary nuclei of about 20 times after a hormonal treatment of 4 days. During the same treatment, the number of β-casein mRNA molecules per cell is markedly augmented, from 40 to 10,000 (Fig. 4A). This clearly indicates that the accumulation of casein mRNA induced by prolactin results from an acceleration of gene transcription and from an increased stability of the mRNA (Teyssot and Houdebine, 1980a). A somewhat similar observation was reported in rat mammary explants (Guyette *et al.*, 1979). As a matter of comparison, the transcription rate of 28 S rRNA genes was evaluated. Prolactin also accelerates this transcription, without simultaneously enhancing the stability of the gene product (Fig. 4B) (Teyssot and Houdebine, 1980a).

Withdrawal of circulating prolactin by injections of bromocriptine into a rabbit during fully established lactation results in a 50% reduction of milk production. This is accompanied by a slow decline of the β-casein gene transcription rate but a rapid and dramatic drop of β-casein mRNA concentration (Fig. 4A) (Teyssot and Houdebine, 1980a). This indicates that the induction and the deinduction of β-casein gene transcription by prolactin are rather slow processes. By contrast, it appears clearly that the stability of the β-casein mRNA is strongly dependent upon the presence of prolactin. The mechanism through which this stabilization is mediated is not known. However, one correlation may be emphasized: the kinetics of β-casein mRNA accumulation (Fig. 4A) is coincident with the formation of the rough endoplasmic reticulum (Houdebine, 1977a) and the deinduction by bromocriptine is accompanied by a simultaneous disappearance of β-casein mRNA and of the endoplasmic reticulum. It is thus conceivable that the binding of the β-casein mRNA on membranes contributes to increase the half-life of the mRNA. It is also worth mentioning that the stabilization of casein mRNA is at most marginally expressed in organ culture (Teyssot and Houdebine, 1981b).

Weaning also leads to a deinduction of the casein gene expression. However, in this situation, the decline of β-casein gene transcription takes place faster than the disappearance of β-casein mRNA (Fig. 4A). Thus, milk accumulation in the mammary gland controls the activity of the genome, through unknown mechanisms.

Deinduction of lactation by prolactin withdrawal or by weaning also leads to a reduction of 28 S rRNA gene expression but obviously via mechanisms different from those involved in the control of casein gene expression (Fig. 4B) (Teyssot and Houdebine, 1980a).

Injections of hydrocortisone acetate with prolactin lead to an acceleration of β-casein gene transcription and a simultaneous accumulation of the mRNA (Fig. 1) (Teyssot and Houdebine, 1981a). In the presence of bromocriptine, the steroid is totally inactive. The glucocorticoid even associated with prolactin is clearly not involved in the control of 28 S rRNA gene expression (Fig. 2). Thus, it appears that glucocorticoids potentiate the action of prolactin on the transcription of casein genes but not most of the other prolactin effects. It is also clear that glucocorticoids support the accumulation of casein mRNA essentially by accelerating

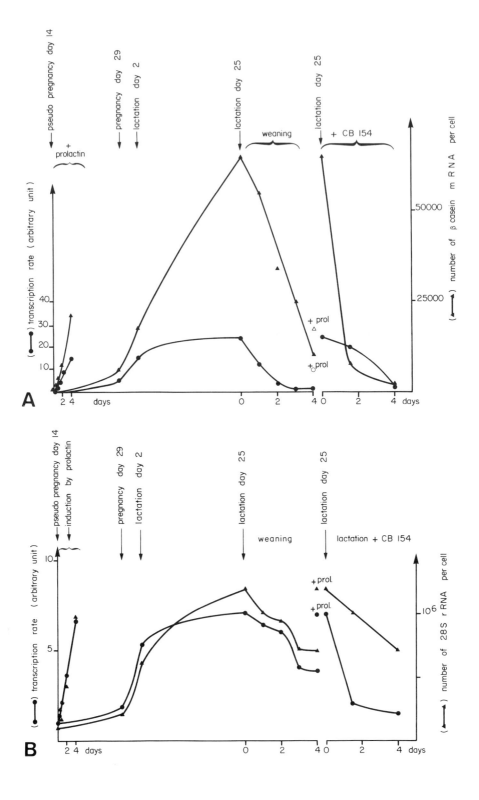

the transcription of the corresponding genes rather than by enhancing the stability of the gene products.

Injections of progesterone with prolactin prevent the acceleration of β-casein and 28 S rRNA gene transcription provoked by prolactin (Figs. 1 and 2) (Teyssot and Houdebine, 1981a). Progesterone thus acts as a general inhibitor of prolactin actions, except for its effect on cell multiplication (Houdebine *et al.*, 1982).

7. Relation between Prolactin Receptor Occupancy and the Hormonal Response

Prolactin receptors have been identified, characterized, and partially purified in the mammary gland (Shiu *et al.*, 1973) and in other tissues (Bergeron *et al.*, 1978). These receptors are located both in the plasma and in the intracellular membranes (Bergeron *et al.*, 1978; Djiane *et al.*, 1981c). The binding of the hormone to its peripheral receptors is assumed to deliver the hormonal message and the hormone–receptor complex is internalized and degraded after a fusion with lysosomes.

7.1. Down-Regulation of Prolactin Receptors

Intravenous injections of prolactin into rabbits result in a rapid and partially reversible occupancy of prolactin receptors in the mammary gland. This phenomenon is accompanied by a progressive and more slowly reversible decline of total number of prolactin receptors (Djiane *et al.*, 1979b). In these experiments the total and free receptors have been discriminated by desaturating the occupied receptors with 5 M MgCl$_2$ (Kelly *et al.*, 1979). An essentially similar effect was observed in cultured mammary explants (Djiane *et al.*, 1981b). The rapid replenishment of prolactin receptors after their down-regulation indicates that the receptor must be a short-living molecule. This fact was more directly established using cycloheximide which induces a dramatic and rapidly reversible decline of prolactin-receptor numbers (Djiane *et al.*, 1979a). Interestingly, this control seems to be exerted mainly on the translation of the receptor mRNA, since inhibitors of transcription do not provoke the disappearance of the receptors (Djiane *et al.*, 1982). In these respects prolactin receptors are subjected to the rules that control the fate of other protein hormone receptors.

Figure 4. Variations of β-casein mRNA and 28 S rRNA accumulation and of transcription rates of the corresponding genes during pregnancy and lactation and under the influence of prolactin. The β-casein mRNA and 28 S rRNA sequences in total mammary RNA and in mercurated RNA synthesized *in vitro* by isolated nuclei were evaluated with the corresponding cDNA probes. The transcription rates refer in all cases to the value found in the noninduced rabbit at day 14 of pseudopregnancy. Prolactin was injected intramuscularly twice per day (100 IU/injection). (A) β-casein gene. (B) 28 S ribosomal RNA genes.

7.2. Up-Regulation of Prolactin Receptors

Independently of the down-regulation process, prolactin receptors are subjected to an up-regulation by the hormone (Djiane *et al.*, 1977). This phenomenon is slow, dependent upon relatively high concentrations of prolactin, and takes place essentially *in vivo*. It is severely inhibited by progesterone injected with prolactin (Djiane and Durand, 1977). Similarly, during pregnancy and more markedly at the initial phase of lactation, the number of prolactin receptors is greatly enhanced. It is tempting to consider that these variations participate with the control of the mammary cell sensitivity towards prolactin. The interpretation of this phenomenon is greatly complicated by the fact that, under the influence of prolactin, the number of receptors in the Golgi apparatus is markedly enhanced due to the overall development of this membrane network. The up-regulation thus corresponds essentially to the presence of the Golgi membranes, and the correlation between cell sensitivity and receptor number remains to be established. This conclusion is reinforced by the fact the prolactin receptors of the Golgi membranes do not generate the prolactin second messenger (see Section 10).

7.3. Relations between Prolactin Concentration in Culture Media, Hormonal Response, and Down-Regulation of Prolactin Receptors

The induction of casein synthesis in mammary explants in culture is a prolactin-dose-dependent phenomenon. In the rabbit, the maximum response is reached at the physiological concentration of 100 ng/ml ovine prolactin, with a slight desensitization at the highest concentrations. This induction is accompanied by an occupancy and a down-regulation of the receptors, as a function of the prolactin concentration (Fig. 5). The antiparallelism of the two phenomena might suggest that they are closely related events.

Experiments designed to block the degradation of the hormone–receptor complex in lysosomes, thus the down-regulation of receptors, indicated that this degradation is not strictly required for prolactin to deliver its message (Houdebine, 1980a; Houdebine and Djiane, 1980b; Djiane *et al.*, 1980). Indeed, NH_4Cl, chloroquine, and other lysosomotropic agents almost totally inhibit the down-regulation of the receptors while not preventing the occupancy of the receptors and the induction of casein synthesis (Fig. 6). The down-regulation of the receptors must therefore be considered as a scavenger process participating in the elimination of the hormonal information rather than as a strictly required event in the mechanism of prolactin action. The fact that prolactin incubated with isolated membranes can generate a second messenger eliciting the hormonal action argues strongly in favor of this interpretation (see Section 10).

The lysosomotropic agents NH_4Cl and chloroquine are generally considered

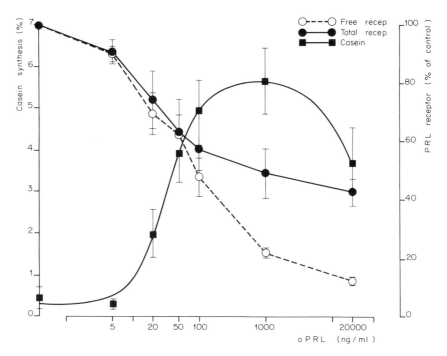

Figure 5. Variations of casein synthesis, prolactin receptor occupancy, and down-regulation in organ culture. Various concentrations of ovine prolactin (NIH-PS 13) were added to 199 medium. After cultures of 24 hr, 50 mg tissue were incubated for 3 hr in Krebs medium containing [^{14}C]-labeled aminoacids. Casein synthesis was then evaluated using anticasein antibodies. Results are expressed as the percentage of labeled proteins immunoprecipitated. The amounts of prolactin receptors were evaluated by the capacity of the membranes to bind [^{125}I]-hGH. Free receptors are titrated by the direct binding to microsomes and total receptors are measured after the desaturation of receptors by 4 M MgCl$_2$.

as not preventing the internalization of receptor–ligand complexes. These drugs are in fact unable to affect the endocytosis of [^{125}I]prolactin when added to culture medium of mammary explants as judged by electron microscopy. The question thus remains of knowing whether the internalization of prolactin is required for prolactin to act. It has been claimed that drugs blocking the cellular transglutaminase, such as bacitracin, dansyl-cadaverine, or methylamine, strongly inhibit the endocytosis of epidermal growth factor (EGF) (Davies *et al.*, 1980). When added to culture media, bacitracin and methylamine do not hamper the induction of casein synthesis. Bacitracin reduces greatly the affinity of prolactin for its receptors while methylamine has essentially a lysosomotropic potency and dansyl-cadaverine rapidly exhibits a cytotoxic effect. None of these drugs appeared to block clearly the endocytosis of the prolactin–receptor complex.

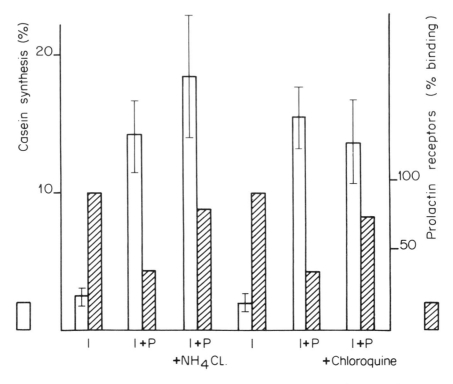

Figure 6.   Effects of NH$_4$Cl and chloroquine on the induction of casein synthesis by prolactin and the down-regulation of prolactin receptors. Insulin (5 μg/ml), was added to the culture medium with or without ovine prolactin (100 ng/ml). After 24 hr of culture, the mammary explants were incubated with labeled aminoacids for the evaluation of casein synthesis (see Fig. 5). Prolactin receptors refer in all cases to total receptors: those which can bind [^{125}I]-hGH after membrane treatment by 4 M MgCl$_2$. NH$_4$Cl and chloroquine when present were at the concentration of 10 mM and 0.1 mM, respectively.

It has also been shown that various drugs inhibiting phospholipase A$_2$ are able to inhibit simultaneously EGF endocytosis (Haigler *et al.*, 1980). These drugs were assayed for their capacity to alter the internalization of prolactin. Quinacrine, phenylglyoxal, chlorpromazine, and *p*-bromophenacyl bromide are more or less efficient, the latter being the most active. None of these drugs is able to prevent the induction of casein synthesis by prolactin, except for quinacrine which is highly cytotoxic. These data argue against the strict requirement of prolactin–receptor complex internalization for prolactin to act. This proposition is of course in agreement with the fact that the prolactin second messenger can be released directly from isolated membranes (see Section 10).

7.4. Persistence of Prolactin Effect after Withdrawal of the Hormone from the Culture Medium

The efficiency of prolactin treatments in pseudopregnant rabbits, in terms of casein-synthesis induction, is dependent upon the amount of prolactin injected and on the route of administration. The same moderate amounts of prolactin are much more efficient in inducing casein synthesis when injected intramuscularly than subcutaneously or intravenously. A comparison between the level of circulating prolactin and the intensity of the response has indicated that a relatively high concentration of prolactin present is permanently required to elicit a high effect. However, the mammary cell can store a significant amount of prolactin information either *in vivo* (Teyssot and Houdebine, 1981a), in organ culture (Houdebine, 1980a), or in cell culture (Servely *et al.*, 1982). Indeed, the addition of prolactin to culture media of isolated cells for only 6 or even 2 hr is sufficient to support a significant hormonal response 1 or 2 days later. The hormonal information has thus been stored in cells. The molecular support of this information is prolactin itself since the persistence of the effect can be totally eliminated by the addition of antiprolactin antibodies or colchicine (Fig. 7) (Servely *et al.*, 1982). After the withdrawal of prolactin from the medium, a very low number of receptors remain occupied by the hormone. This indicates that only a small number of occupied receptors is sufficient to ensure the hormonal response, and suggests that the intensity of the response and the occupancy of the receptors (Fig. 5) are not strictly correlated events. In addition, this demonstrates that an amplification of the prolactin message must take place at the receptor level, and that many more than one prolactin second messenger may be produced per occupied receptor.

8. Action of Antiprolactin Receptor Antibodies

In order to study the role of prolactin receptors in the transmission of the hormonal message, the receptor was partially purified using affinity chromatography essentially according to Shiu and Friesen (1974). The receptor fraction was injected into guinea pigs, sheep, and goats. In all cases, antibodies were obtained which specifically inhibit the binding of [125I]prolactin to its receptors. When added to culture media of explants or isolated cells with prolactin, the antireceptor-containing serum prevented the ability of prolactin to induce casein synthesis. These observations confirm previous work carried out by Shiu and Friesen (1976). When added alone, in the absence of prolactin, the antireceptor containing serum exhibited a potent prolactinlike activity (Fig. 8) (Djiane *et al.*, 1981a). This effect was obtained with all the sera tested and with the immunoglobulin fractions derived from immunized but not from nonimmunized ani-

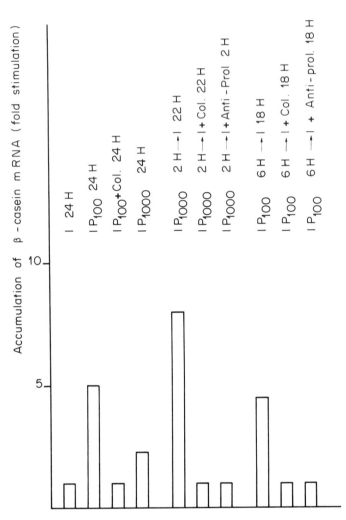

Figure 7.   Effect of prolactin withdrawal from culture media on the induction of β-casein mRNA accumulation. Mammary isolated epithelial cells were cultured in the presence of serum for about 1 week. The serum was then withdrawn and the cells kept for 2 days in MEM medium containing insulin. After these treatments, ovine prolactin (100 ng or 1000 ng/ml as stated) was added to this culture medium for 2, 6, or 24 hr. After 2 or 6 hr, the hormone was withdrawn and cultures were pursued for 22 or 18 hr, respectively, in the presence or not of colchicine (3 μM) or horse antiovine prolactin (1 mg/ml). β-casein mRNA concentrations were evaluated in all cases 24 hr after the initial addition of prolactin using a β-casein cDNA probe. In the presence of insulin alone the number of β-casein mRNA molecules per cell was seven.

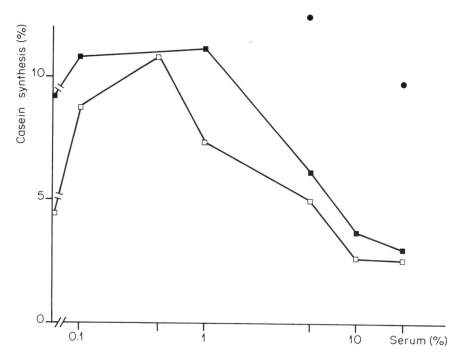

Figure 8.   Action of anti-prolactin-receptor antibodies on casein synthesis. Serum from a sheep immunized with partially purified prolactin receptors was added to the culture medium of mammary explants containing insulin and cortisol, in the presence or in the absence of prolactin (100 ng/ml). After 24 hr culture casein synthesis was evaluated using anticasein antibodies. Results are expressed as the percentage of labeled proteins immunoprecipitated. (□) Without prolactin, (■) with prolactin, (●) with prolactin and in the presence of the serum of a nonimmunized sheep.

mals. Interestingly, the immunoglobulin fraction was also capable of stimulating DNA synthesis in cultured mammary explants. The induction of casein mRNA accumulation by the antiprolactin receptor antibodies was amplified by cortisol, unaltered by lysosomotropic agents, and inhibited by colchicine. Injections of the antireceptor immunoglobulins into pseudopregnant rabbits were also able to induce the synthesis of milk clearly visible in the mammary gland. This induction was coincident with the accumulation of casein mRNA and the effects of the antibody were inhibited by the simultaneous injection of progesterone. The antiprolactin antibodies thus appear to mimic several of the essential prolactin actions on the mammary cell. The main conclusion which can be drawn from these observations is that the prolactin molecule is most likely not required beyond its binding to receptors and that an intracellular relay is formed as soon as a specific ligand is bound to the receptor. Results in Section 10 are compatible with this interpretation.

9. Role of Tubulin-Containing Structures in the Transmission of the Prolactin Message to Casein Genes

It has been proposed that the mobility of plasma-membrane components and possibly of hormone receptors is under the control of the cytoskeleton (Edelman, 1976). In order to evaluate the possible involvement of the cytoskeleton in the transmission of the prolactin message, colchicine and related drugs were added to the culture medium of explants and of isolated cells. Colchicine, vinblastin, podophyllotoxin colcemid, and nocodazole prevented prolactin from inducing casein and DNA synthesis (Houdebine *et al.*, 1979; Houdebine and Djiane, 1980b; Houdebine, 1980a), whereas three colchicine analogs not binding tubulin (lumicolchicine, trimethylcolchicinic acid, and colchicein) were essentially inactive (Houdebine 1980a, 1981b). When injected with prolactin into pseudopregnant rabbits, colchicine prevented the accumulation of β-casein mRNA and the acceleration of the transcription of the corresponding gene (Fig. 9) (Teyssot and

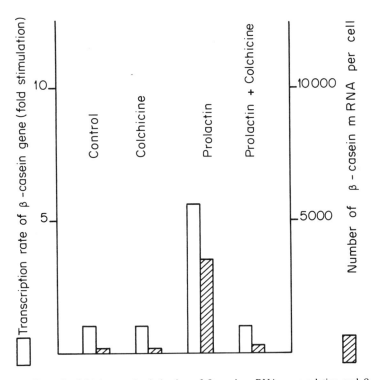

Figure 9.   Effect of colchicine on the induction of β-casein mRNA accumulation and β-casein transcription by prolactin. Prolactin and colchicine were injected intramuscularly (100 IU/injection) and subcutaneously (4 mg/injection) respectively, twice daily for 2 days. β-casein sequences in total mammary RNA and in mercurated RNA synthesized by isolated nuclei were evaluated with a β-casein cDNA probe.

Houdebine, 1980b). Surprisingly, the drug did not inhibit the accumulation of 28 S rRNA and the stimulation of casein synthesis (see Section 5) (Teyssot and Houdebine, 1980b, 1981b). These data indicate that the tubulin-binding drugs inhibit only part of the prolactin message. It is striking that several prolactin actions that take place *in vivo* but not in organ culture (thus probably not mediated through the sole direct hormonal action on the mammary cell) are neither inhibited

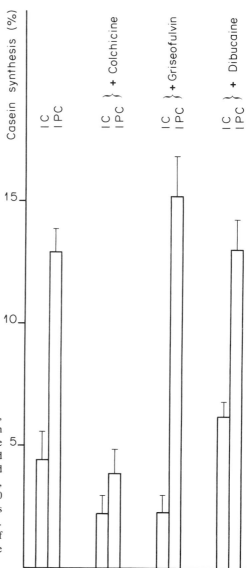

Figure 10. Effects of colchicine, griseofulvin, and dibucaine on the induction of casein synthesis by prolactin. Culture media of mammary explants contained insulin (5 μg/ml), cortisol (500 ng/ml), and when stated, ovine prolactin (100 ng/ml), colchicine (1 μM), and griseofulvin (100 μM). After 24-hr culture, casein synthesis was estimated using anticasein antibodies. Results are expressed as the percentage of labeled proteins precipitated by the antibodies.

by colchicine nor amplified by glucocorticoids (rRNA synthesis, casein mRNA stabilization, casein mRNA translation) (Teyssot and Houdebine, 1981b).

These data might lead to the conclusion that the microtubules and more generally the cytoskeleton are involved in the transmission of the prolactin message. To check that point further, several drugs disrupting the cytoskeleton but not binding tubulin at the colchicine site were added to the culture medium of mammary explants together with prolactin. The destabilization of microfilaments by cytochalasin B was unable to affect significantly the prolactin-dependent accumulation of β-casein mRNA (Houdebine and Djiane, 1980b). Neither did griseofulvin which is considered to disrupt microtubules (Weber et al., 1976; Roobol et al., 1977; David-Pfeuty et al., 1979) inhibit prolactin action (Fig. 10) (Houdebine et al., 1981a). Similarly, local anesthetics which have been demonstrated to destabilize the link between the cytoskeleton and plasma membrane were unable to prevent prolactin action (Fig. 10) (Houdebine et al., 1981b). In addition, these drugs did not inhibit the capacity of prolactin to stimulate DNA synthesis and to down-regulate its own receptor. A real involvement of the cytoskeleton in the transmission of the prolactin message in the mammary cell is therefore by no means evident. It is rather tubulin itself or other colchicine-binding molecules located in the plasma membrane which play a role in the mechanism of prolactin action. The existence of such binding entities in mammary membrane argues in favor of such an interpretation (Houdebine et al., 1981a). Still more convincing is the fact that the prolactin second messenger can be released from purified plasma membranes and that this release is blocked by colchicine (see Section 10).

10. Identification of a Possible Second Messenger for Prolactin

Prolactin is one of the hormones for which no intracellular relay has been found so far. It is generally admitted that neither cAMP or cGMP is directly involved in the transmission of the prolactin message. It has been proposed that prolactin itself associated or not with its receptor might migrate to nuclei and carry the information (Nolin and Bogdanove, 1980). The amount of prolactin transfered to nucleus is in fact rather low and hardly capable of accounting for a hormonal action (Houdebine, 1980b,c). Other possible intracellular relays such as polyamines (Houdebine et al., 1978a), potassium ions (Houdebine and Djiane, 1980a), calcium ions (Houdebine, 1981a), or prostaglandins (Houdebine and Lacroix, 1980) are most likely not the prolactin second messengers. It is also clear that calmodulin is not strictly involved since trifluoperazine added to culture media was without any effect on the induction of casein synthesis by prolactin (not shown). It seems therefore that a nonclassic mechanism is to be considered to account for the transmission of the prolactin message to genes.

The possibility remained that the intracellular relay is still released by prolactin directly from the membrane and that this relay activates directly casein gene transcription after having been transferred to the nucleus. This hypothesis was subjected to the following experimentation. Mammary microsomes were incubated with or without prolactin for 1 hr at room temperature and the resulting soluble fractions were saved after pelleting the membranes. This soluble fraction was incubated with mammary nuclei isolated from pseudopregnant noninduced rabbits or from lactating rabbits deinduced by bromocriptine injections. The isolated nuclei were incubated in the presence of Hg CTP and the mercurated RNA were assayed for their content in β-casein mRNA sequences using the corresponding labeled cDNA probe, as described in Section 6: the transcription of β-casein genes was enhanced 5–10 times in isolated nuclei incubated in the presence of the membrane supernatant obtained with prolactin. The control supernatant obtained without prolactin was totally devoid of stimulating activity. Interestingly, the membrane supernatants were active exclusively when incubation was carried out in the presence of lactogenic hormones (Fig. 11). Prolactin added directly to nuclei without having been previously incubated with the membranes was totally inactive. The factor was also generated by membranes isolated from various prolactin-receptor-containing tissues but not from those devoid of these receptors (Fig. 12). The factor is not a general transcription stimulator since the total RNA synthesis in incubated nuclei and the activity of RNA polymerase II were essentially unaltered by the membrane supernatants (Teyssot *et al.*, 1981a). On the other hand, it was unable to induce the transcription of β-casein genes in rabbit liver and reticulocyte nuclei, or to stimulate the transcription of housekeeping and ribosomal rRNA genes in mammary nuclei and the transcription of globin genes in reticulocyte nuclei (Teyssot *et al.*, 1981a). This factor has thus several of the essential criteria to be considered as specific and appears a good candidate for the prolactin second messenger carrying the hormonal information to prolactin-sensitive genes.

Further experiments demonstrated that the factor is released from plasma but not from Golgi membranes, indicating that only the receptors located in the former are strictly involved in prolactin action. The prolactin relay was not generated by membranes when incubation was carried out in the presence of colchicine. Interestingly, the factor was generated by serum containing anti-prolactin-receptor antibodies (Fig. 13). The release of the factor was dependent upon the concentration of prolactin in the incubate and it took place at 25°C and 37°C but not at 4°C. The active entity in the membrane supernatant is a small-size compound eluted similarly to a tripeptide from a Sephadex G-25 column and it is inactivated by trypsin, indicating that it contains a polypeptide chain (Teyssot *et al.*, 1982). The properties of the prolactin intracellular relay so far identified are strikingly reminiscent of the insulin second messenger activating glycogen synthetase (Larner *et al.*, 1979) and pyruvate dehydrogenase (Seals and Jarett, 1980; Seals and Czech, 1980; Kiechle *et al.*, 1981). Hence, the

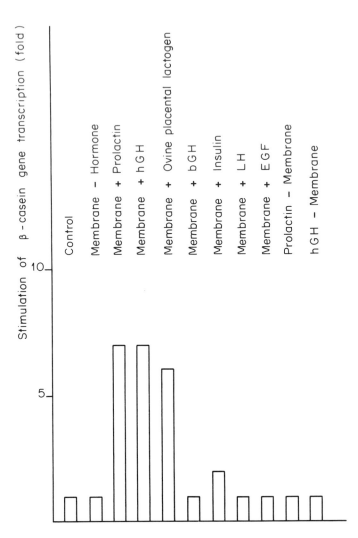

Figure 11.   Effects of factors released from membranes on the transcription rate of β-casein gene. Mammary microsomes were incubated with various lactogenic hormones (prolactin, hGH, and ovine placental lactogen) and nonlactogenic hormones (hGH, insulin, LH, and EGF). The membranes' supernatants were saved and incubated with mammary nuclei isolated from lactating rabbits treated for 4 days with Bromocriptine in the presence of Hg CTP. The concentration of β-casein mRNA sequences in the mercurated RNA retained by and eluted from SH-Sepharose was estimated using a β-casein cDNA probe. Results refer to the control value obtained in the absence of membrane supernatants.

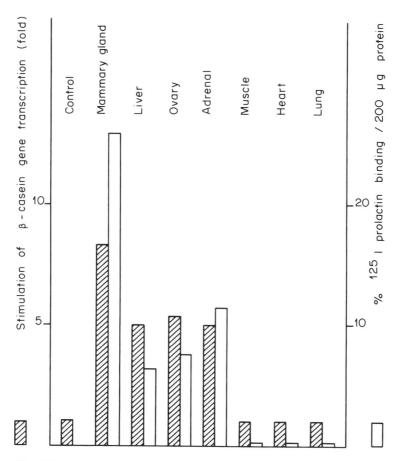

Figure 12. Effects of membranes' supernatants from various tissues on the transcription rate of β-casein genes. Microsomes from various tissues were incubated for 1 hr at room temperature with ovine prolactin (5 μg/ml). The membranes' supernatants were incubated with mammary nuclei as in Fig. 11. β-casein gene transcription refers to the value obtained with nuclei incubated in the absence of membranes' supernatant. The presence of prolactin receptors in each membrane preparation was estimated from their capacity to bind [^{125}I]prolactin.

mechanism of prolactin action proposed here might be of general occurrence for polypeptide hormones.

11. Conclusions

The experimental data reported here clearly indicate that the control of casein gene expression is exerted at multiple levels: gene transcription, mRNA

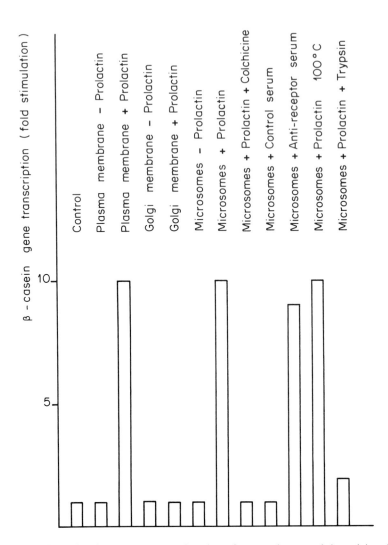

Figure 13.   Effects of various treatments on the release from membranes and the activity of the factor which stimulates β-casein gene transcription in isolated nuclei. Purified plasma, Golgi membranes, and crude microsomes from rabbit liver were incubated with and without prolactin and the supernatants were added to isolated nuclei. Microsomes were incubated with prolactin in the presence or in the absence of colchicine (5 μM) and the supernatants were added to isolated nuclei. Microsomes were incubated with sheep serum (10%) immunized or not with prolactin receptors and the supernatants were added to isolated nuclei. The membrane supernatants obtained with prolactin were heated at 100°C for 10 min or treated by trypsin (1 μg/ml) for 15 min before being added to isolated nuclei. In all cases, the transcription rate of β-casein genes was quantitated as described in Figure 11.

stabilization, and mRNA translation. Prolactin is involved in each of these steps and acts as an inducer. Glucocorticoids are potent amplifiers of prolactin action. However, a striking feature concerning these steroids is that they essentially amplify the transcription of casein genes. This effect seems sufficient to account for the resulting accumulation of casein mRNA which in turn leads to an enhancement of casein synthesis. There is no evidence that glucocorticoids participate in a specific stabilization of casein mRNAs. In these respects, glucocorticoids work in the mammary cell essentially as in other cell types: they potentiate the action of other specific inducers on specific genes. Another typical aspect is the relative sluggishness of prolactin action to modify casein gene transcription: the transcription is accelerated and deinduced rather slowly. On the other hand, the specific stabilization of casein mRNA is an essential part of the mechanism leading to mRNA accumulation. Prolactin is also greatly involved in the control of casein mRNA translation. Progesterone attenuates essentially all prolactin actions, except for cell multiplication.

The full expression of casein genes thus appears to be the result of a set of amplifications: cell multiplication, accumulation of casein mRNA, cellular hypertrophy, induction of casein mRNA translation. These amplifications reflect the transient existence of the secreting mammary cell and its requirement for a high metabolic activity. In contrast with the transcription of casein genes, all these events do not operate significantly in cultured cells, suggesting that some of them are mediated through unknown relays in the organism. Cultured explants or cells accurately reflect prolactin actions on casein mRNA accumulation, casein synthesis, and cell multiplication but only marginally casein mRNA stabilization and translation and cellular hypertrophy. It suggests that the effects obtained in cultures, thus certainly resulting from the direct action of prolactin, are those amplified by glucocorticoids and inhibited by tubulin-binding drugs. Conversely, progesterone proved to be very poorly efficient when added to culture media (Teyssot and Houdebine, 1981a; Houdebine *et al.*, 1982). This suggests that the inhibitory action of this steroid is exerted essentially on the indirect effects of prolactin. All these conclusions are summarized schematically in Fig. 14. Little is known about the mechanisms involved in the indirect actions of prolactin. These indirect actions are assumed to play an essential role in providing milk precursors and possibly prolactin-mediated hormonal information to the mammary cells. It is conceivable that this control is ensured by several synergistic hormones such as growth hormone, thyroid hormones, glucocorticoids, etc. In the rabbit, this role might be essentially played by prolactin itself, rendering this species exceptionally sensitive to the hormone for the induction of mammary gland development and secretory activity.

One major step of the direct action of prolactin on the mammary cell (and possibly on other cells) seems to have been elucidated by the discovery of a factor released from the membrane and stimulating casein gene transcription in isolated nuclei. The specificity of action of this factor strongly suggests that it

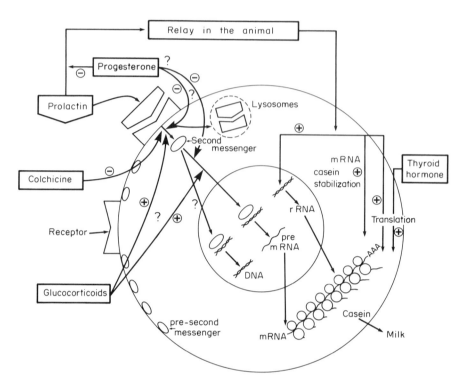

Figure 14.   Schematic representation of the hormonal mechanisms controlling casein gene expression. Presecond messenger for prolactin is present in the plasma membrane and the active relay is
released when prolactin or antiprolactin-receptor antibodies bind to the receptor. This release is
blocked by colchicine. The intracellular relay moves to the nucleus where it activates the prolactin-
sensitive genes and possibly DNA replication. It is not known whether glucocorticoids favor the
generation of the relay or its action on the nucleus. Independently from this mechanism, the prolactin–receptor complex is internalized and degraded in lysosomes. Some of the prolactin actions
are most likely expressed via relays in the animal. These actions are strongly inhibited by progesterone
which does not seem to act directly on the mammary cell. The mechanisms leading to the generation
of the second messenger and to its action on the nucleus remain unknown.

is the prolactin second messenger carrying the hormonal information to genes.
The fact that the factor is produced by several prolactin-receptor-containing
membranes indicates that a new mechanism for polypeptide hormone has perhaps
been discovered. This assumption is of course greatly reinforced by the recent
identification of the second messenger for insulin (Larner *et al.,* 1979; Seals and
Jarett, 1980; Seals and Czech, 1980; Kiechle *et al.,* 1981). The excellent correlation between the effects of hormones, colchicine, and antiprolactin-receptor
antibodies on the cell and on the membranes strongly argue in favor of the
conclusion that the experiments depicted here really reflect a biological event.

Experiments are in progress in our laboratories to isolate the factor and to determine through which mechanisms it is generated from the membrane and activates casein gene transcription.

Note Added in Proof

Since the manuscript was first submitted, the following observations have been made. The antiprolactin receptor mimicked prolactin action in hepatocytes for the up-regulation of the receptor (Rosa *et al.*, 1982, *Biochem. Biophys. Res. Comm.*, **106**:243–249). Monovalent antibodies (Fab) did not mimic prolactin action although at high concentration they kept their capacity to inhibit prolactin binding and action. When added to the culture medium of isolated mammary cells, various fractions containing the prolactin intracellular relay mimicked prolactin action for the accumulation of β-casein mRNA (Servely *et al.*, 1982, *FEBS Lett.* **148**:242–246). The stimulatory effect of the prolactin relay on isolated nuclei was observed for αs_1- and β-caseins and for α-lactalbumin, using the corresponding plasmids as probes. The stimulation of β-casein gene transcription by prolactin relay was totally abolished by several phosphatase inhibitors, suggesting that the mediator works via a dephosphorylation of nuclear proteins (Houdebine *et al.*, *Prolactine, Neurotransmission et Fertilité* [H. Clauser and J. P. Gautray, eds.], Masson, pp. 53–65). This constitutes an additional analogy with the insulin intracellular relay (Houdebine *et al.*, *Reprod. Nutr. Develop.*, in press).

ACKNOWLEDGMENTS. This work has been carried out with the technical assistance of Mrs. Puissant, L. Belair, M. L. Fontaine, and Mr. H. Grabowski. The authors wish to thank Professor H. Clauser for helpful encouragement and criticism. This work was supported by the financial help of the Centre National de la Recherche Scientifique, the Institut National de la Santé et de la Recherche Medicale and the Délégation Générale pour la Recherche Scientifique et Technique.

REFERENCES

Assairi, L., Delouis, C., Gaye, P., Houdebine, L. M., Ollivier-Bousquet, M., and Denamur, R., 1974, Inhibition by progesterone of the lactogenic effect of prolactin in the pseudopregnant rabbit, *Biochem. J.* **144**:245–252.
Bergeron, J. J. M., Posner, B. J., Josefsberg, S., and Silkstrom, R., 1978, Intracellular polypeptide hormone receptors: The demonstration of specific binding sites for insulin and human growth hormone in Golgi fractions isolated from the liver of female rats, *J. Biol. Chem.* **253**:4058–4066.
Burditt, L. J., Parker, D., Craig, R. K., Getova, T., and Campbell, P. N., 1981, Differential expression of α-lactalbumin and casein genes during the onset of lactation in the guinea-pig mammary gland, *Biochem. J.* **194**:999–1006.

David-Pfeuty, T., Simon, C., and Pantaloni, D., 1979, Effect of antimitotic drugs on tubulin GTPase activity and self assembly, *J. Biol. Chem.* **254**:11696–11701.

Davies, P. J. A., Davies, D. R., Levitzki, A., Maxfield, F. R., Milhaud, P., Willingham, M. C., and Pastan, I. H., 1980, Transglutaminase is essential in receptor mediated endocytosis of α₂-macroglobulin and polypeptide hormones, *Nature* **283**:162–167.

Denamur, R., 1971, Hormonal control of lactogenesis, *J. Dairy Sci.* **38**:237–264.

Devinoy, E., and Houdebine, L. M., 1977, Effects of glucocorticoids on casein gene expression in the rabbit, *Eur. J. Biochem.* **75**:411–416.

Devinoy, E., Houdebine, L. M., and Delouis, C., 1978, Role of prolactin and glucocorticoids in the expression of casein genes in rabbit mammary gland organ culture. Quantification of casein mRNA, *Biochim. Biophys. Acta* **517**:360–366.

Devinoy, E., Houdebine, L. M., and Ollivier-Bousquet, M., 1979, Role of glucocorticoids and progesterone in the development of rough endoplasmic reticulum involved in casein biosynthesis, *Biochimie* **61**:453–461.

Djiane, J., and Durand, P., 1977, Prolactin-progesterone antagonism in self regulation of prolactin receptors in the mammary gland, *Nature,* **266**:641–643.

Djiane, J., Durand, P., and Kelly, P. A., 1977, Evolution of prolactin-receptors in rabbit mammary gland during pregnancy and lactation, *Endocrinology* **100**:1348–1356.

Djiane, J., Delouis, C., and Kelly, P. A., 1979a, Prolactin receptors in organ culture of rabbit mammary gland: Effect of cycloheximide and prolactin, *Proc. Soc. Exp. Biol. Med.* **162**:342–347.

Djiane, J., Kelly, P. A., and Clauser, H., 1979b, Rapid down-regulation of prolactin receptors in mammary gland and liver, *Biochim. Biophys. Res. Comm.* **90**:1371–1378.

Djiane, J., Kelly, P. A., and Houdebine, L. M., 1980, Effects of lysosomotropic agents, cytochalasin B and colchicine on the down-regulation of prolactin receptors in mammary gland explants, *Mol. Cell. Endocrinol.* **18**:87–98.

Djiane, J., Houdebine, L. M., and Kelly, P. A., 1981a, Prolactin-like activity of anti-prolactin receptor antibodies on casein and DNA synthesis in the mammary gland, *Proc. Natl. Acad. Sci USA* **78**:7445–7448.

Djiane, J., Houdebine, L. M., and Kelly, P. A., 1981b, Correlation between prolactin-receptor interaction; down-regulation of receptors and stimulation of casein and DNA biosynthesis in rabbit mammary gland explants, *Endocrinology* **110**:791–795.

Djiane, J., Houdebine, L. M., and Kelly, P. A., 1981c, Down-regulation of prolactin receptors in rabbit mammary gland: Differential subcellular localization, *Proc. Soc. Exp. Biol. Med.* **168**:*378–381*.

Djiane, J., Delouis, C., and Kelly, P. A., 1982, Prolactin receptor turn-over in pseudopregnant rabbit mammary gland explants, *Mol. Cell. Endocrinol.* **25**:163–170.

Edelman, G. M., 1976, Surface modulation in cell recognition and cell growth, *Science* **192**:218–226.

Fevre, J., and Houdebine, L. M., 1978, Glucocorticoids and mammary gland development: Mammary cell multiplication and hypertrophy in rabbit, *Ann. Biol. Anim. Biochem. Biophys.* **18**:1325–1331.

Ganguly, R., Mehta, N. M., Ganguly, N., and Banerjee, M. R., 1979, Glucocorticoid modulation of casein gene transcription in mouse mammary gland, *Proc. Natl. Acad. Sci. USA* **76**:6466–6470.

Guyette, W. A., Matusik, R. J., and Rosen, J. M., 1979, Prolactin-mediated transcriptional and post-transcriptional control of casein gene expression, *Cell* **17**:1013–1023.

Haigler, H. T., Willinghan, M. C., and Pastan, I., 1980, Inhibitors of ¹²⁵I-epidermal growth factor internalization, *Biochem. Biophys. Res. Commun.* **94**:630–637.

Houdebine, L. M., 1976, Effects of prolactin and progesterone on expression of casein genes. Titration of casein mRNA by hybridization with complementary DNA, *Eur. J. Biochem.* **68**:219–225.

Houdebine, L. M., 1977a, Distribution of casein mRNA between free and membrane-bound polysomes during induction of lactogenesis in the rabbit, *Mol. Cell. Endocrinol.* **7**:125–135.

Houdebine, L. M., 1977b, Role of prolactin in the expression of casein genes in the virgin rabbit, *Cell Differ.* **8**:49–59.

Houdebine, L. M., 1980a, Effect of various lysosomotropic agents and microtubule disrupting drugs on the lactogenic and mammogenic action of prolactin, *Eur. J. Cell. Biol.* **22:**755–760.

Houdebine, L. M., 1980b, The control of casein gene expression by prolactin and its modulator, in: *Central and Peripheral Regulation of Prolactin* (R. M. MacLeod and U. Scapagnini, eds.), pp. 189–206, Raven Press, New York.

Houdebine, L. M., 1980c, Role of prolactin, glucocorticoids and progesterone in the control of casein gene expression, in: *Hormone and Cell Regulation* (J. Dumont and J. Nunez, eds.), pp. 175–196, Elsevier/North Holland Biomedical Press, Amsterdam.

Houdebine, L. M., 1981a, Rôle du calcium dans l'induction de la synthèse des caséines par la prolactine dans la glande mammaire de lapine, *Biol. Cell.* **40:**129–134.

Houdebine, L. M., 1981b, Effet de la colchiceine sur l'action lactogène et mammogène de la prolactine, *Biol. Cell.* **40:**135–138.

Houdebine, L. M., and Djiane, J., 1980a, Effet de l'oubaine sur l'action lactogène de la prolactine et sur le niveau des récepteurs prolactiniques mammaires, *Biochimie* **62:**433–440.

Houdebine, L. M., and Djiane, J., 1980b, Effects of lysosomotropic agents and of microfilaments and microtubule disrupting drugs on the activation of casein gene expression by prolactin in the mammary gland, *Mol. Cell. Endocrinol.* **17:**1–15.

Houdebine, L. M., and Gaye, P., 1975, Regulation of casein synthesis in the rabbit mammary gland. Titration of mRNA activity for casein under prolactin and progesterone treatments, *Mol. Cell. Endocrinol.* **3:**37–55.

Houdebine, L. M., and Lacroix, M. C., 1980, Effet de l'indométhacine sur l'action lactogène de la prolactine, *Biochimie* **62:**441–444.

Houdebine, L. M., Devinoy, E., and Delouis, C., 1978a, Role of spermidine in casein gene expression in the rabbit, *Biochimie* **60:**735–741.

Houdebine, L. M., Delouis, C., and Devinoy, E., 1978b, Post-transcriptional stimulation of casein synthesis by thyroid hormone, *Biochimie* **80:**809–812.

Houdebine, L. M., Djiane, J., and Clauser, H., 1979, Rôle des lysosomes, des microtubules et des microfilaments dans le mécanisme d'action lactogène de la prolactine sur la glande mammaire de lapine, *C. R. Acad. Sci. Paris* **289:**679–682.

Houdebine, L. M., Ollivier-Bousquet, M., and Djiane, J., 1981a, Rôle des protéines membranaires liant la colchicine dans la transmission du message prolactinique aux gènes des caséines dans la glande mammaire de lapine, *Biochimie* **64:**21–28.

Houdebine, L. M., Djiane, J. and Ollivier-Bousquet, M., 1981b, Effects of local anesthetics on the transmission of prolactin message to casein genes and on the down-regulation of prolactin receptor, *Biol. Cell.* **41:**231–234.

Houdebine, L. M., Teyssot, B., Devinoy, E., Ollivier-Bousquet, M., Djiane, J., Kelly, P. A., Delouis, C., Kann, G., and Fèvre, J., 1982, Role of progesterone in the development and the activity of mammary gland, in: *Progesterone and Progestins* (C. W. Bardin, P. Mauvais-Jarvis, and E. Milgrom, eds.), pp. 301–323, Raven Press, New York.

Kelly, P. A., Leblanc, G., and Djiane, J., 1979, Estimation of total prolactin binding sites after *in vitro* desaturation, *Endocrinology* **104:**1631–1638.

Kiechle, F. L., Jarett, L., Kotagal, N., and Popp, D. A., 1981, Partial purification from rat adipocyte plasma membranes of a chemical mediator which stimulates the action of insulin on pyruvate dehydrogenase, *J. Biol. Chem.* **256:**2945–2951.

Larner, J., Galasko, G., Cheng, K., Depaoliroach, A. A., Huang, L., Daggy, P., and Kellog, J., 1979, Generation by insulin of a chemical mediator that controls protein phosphorylation and dephosphorylation, *Science* **206:**1408–1411.

Matusik, R. J., and Rosen, J. M., 1978, Prolactin induction of casein mRNA in organ culture, *J. Biol. Chem.* **253:**2343–2347.

Mehta, N. M., Ganguly, N., Ganguly, R., and Banerjee, M. R., 1980, Hormonal modulation of the casein gene expression in a mammogenesis-lactogenesis culture model of the whole mammary gland of the mouse, *J. Biol. Chem.* **255:**4430–4434.

Mercier, J. C., and Gaye, P., 1980, Study of secretory lactoproteins: Primary structures of the signals and enzymatic processing, *Ann. NY Acad. Sci.* **343:**232–251.

Mercier, J. C., Grosclaude, F., and Ribadeau-Dumas, B., 1972, Primary structure of bovine caseins, A review, *Milchwissenschaft* **27:**402–408.

Nakhasi, H. L., and Qasba, P. K., 1979, Quantitation of milk proteins and their mRNAs in rat mammary gland at various stages of gestation and lactation, *J. Biol. Chem* **254:**6016–6025.

Nicolson, G. L., Smith, J. R., and Poste, G., 1976, Effects of local anesthetics on cell morphology and membrane associated cytoskeletal organization in BALB/3T3 cells, *J. Cell. Biol.* **68:**395–402.

Nolin, J. M., and Bogdanove, E. M., 1980, Effects of estrogen on prolactin incorporation by lutein and milk secretory cells and on pituitary prolactin secretion in the post-partum rat: Correlations with target cell responsiveness to prolactin, *Biol. Reprod.* **22:**393–416.

Qasba, P. K., Dandekar, A. M., Sobiech, K. A., Nakhasi, H. L., Devinoy, E., Horn, T., Losonczy, I., and Siegel, M., 1982, Milk protein gene expression in rat mammary gland, in: *Critical Reviews in Food Sciences and Nutrition* CRC, Boca Raton, Florida, Vol. 16, pp. 164–189.

Roobol, A., Gull, K., and Pogson, C. I., 1977, Evidence that griseofulvin binds to a microtubule associated protein, *FEBS Letts* **75:**149–153.

Rosen, J. M., Woo, S. L. C., and Comstock, J. P., 1975, Regulation of casein messenger RNA during the development of the rat mammary gland, *Biochemistry* **14:**2895–2903.

Rosen, J. M., Supowit, S. C., Gupta, P., Yu, L. Y., and Hobbs, A. A., 1981, Regulation of casein gene expression in hormone-dependent mammary cancer, in: *Bradbury Report 8: Hormones and Breast Cancer* (M. Pike, P. K. Siiteri, and C. W. Welsch, eds.), pp. 397–424, Cold Spring Harbor Press, New York.

Seals, J. R., and Czech, M. P., 1980, Evidence that insulin activates an intrinsic plasma membrane protease in generating a second chemical mediator, *J. Biol. Chem.* **255:***6529–6531.*

Seals, J. R., and Jarett, L., 1980, Activation of pyruvate dehydrogenase by direct addition of insulin to an isolated plasma membrane/mitochondria mixture: Evidence for generation of insulin's second messenger in a subcellular system, *Proc. Natl. Acad. Sci. USA* **77:**77–81.

Servely, J. L., Teyssot, B., Houdebine, L. M., Delouis, C., and Djiane, J., 1981, Evidence that the activation of casein gene expression in the rabbit mammary gland can be elicited by a low amount of prolactin firmly retained on its receptors, *Biochimie* **64:**133–140.

Servely, J. L., N'Gueme Emana, M., Houdebine, L. M., Djiane, J., Delouis, C., and Kelly, P. A., 1983, Comparative measurement of the lactogenic activity of ovine placental lactogen in rabbit and ewe mammary gland. *Gen. Comp. Endocrinol.* (in press).

Shiu, R. P. C., and Friesen, H. G., 1974, Solubilization and purification of a prolactin receptor from the rabbit mammary gland, *J. Biol. Chem.* **249:**7902–7911.

Shiu, R. P. C., and Friesen, H. G., 1976, Blockade of prolactin action by an antiserum to its receptors, *Science* **192:**259–261.

Shuster, R. C., Houdebine, L. M., and Gaye, P., 1976, Studies on the synthesis of casein messenger RNA during pregnancy in the rabbit, *Eur. J. Biochem.* **71:**193–199.

Takemoto, T., Nagamatsu, Y., and Oka, T., 1980, Casein and α-lactalbumin messenger RNA during the development of mouse mammary gland, *Dev. Biol.* **78:**247–257.

Teyssot, B., and Houdebine, L. M., 1980a, Role of prolactin in the transcription of β-casein and 28S ribosomal genes in the rabbit mammary gland, *Eur. J. Biochem.* **110:**263–272.

Teyssot, B., and Houdebine, L. M., 1980b, Effects of colchicine on the transcription rate of β-casein and 28S ribosomal RNA genes in the rabbit mammary gland, *Biochem. Biophys. Res. Commun.* **97:**463–473.

Teyssot, B., and Houdebine, L. M., 1981a, Role of progesterone and glucocorticoids in the transcription of the β-casein and 28S-RNA genes in the rabbit mammary gland, *Eur. J. Biochem* **114:**597–608.

Teyssot, B., and Houdebine, L. M., 1981b, Induction of casein synthesis by prolactin and inhibition by progesterone in the pseudopregnant rabbit treated by colchicine without any simultaneous variations of casein mRNA concentration, *Eur. J. Biochem.* **117:**563–568.

Teyssot, B., Houdebine, L. M., and Djiane, J., 1981a, Prolactin induces release of a factor from membranes capable of stimulating β-casein gene transcription in isolated mammary cell nuclei, *Proc. Natl. Acad. Sci. USA* **78**:6729–6733.

Teyssot, B., Servely, J. L., Delouis, C., and Houdebine, L. M., 1981b, Control of casein gene expression in isolated cultured rabbit epithelial mammary cells, *Mol. Cell. Endocrinol.* **23**:33–48.

Teyssot, B., Djiane, J., Kelly, P. A., and Houdebine, L. M., 1982, Identification of the putative prolactin second messenger activating β-casein gene transcription, *Biol. Cell* **43**:81–88.

Vina, J., Puertes, I. R., Saez, G. T., and Vina, J. R., 1981, Role of prolactin in aminoacid uptake by the lactating mammary gland of the rat, *FEBS Lett.* **126**:250–252.

Vonderhaar, B. K., Smith, G. H., Pauley, R. J., Rosen, J. M., and Topper, Y. J., 1978, A difference between mammary epithelial cells from mature virgin and primiparous mice, *Cancer Res.* **38**:4059–4065.

Weber, K., Wehland, J., and Herzog, W., 1976, Griseofulvin interacts with microtubules both *in vivo* and *in vitro*, *J. Mol. Biol.* **102**:817–829.

6

Sequential Regulation of Gene Expression by Estrogen in the Developing Rat Uterus

Alvin M. Kaye

1. Introduction

The study of gene expression in eukaryotic systems has, to date, followed two contrasting approaches: RNA–DNA hybridization surveys of the range of newly synthesized mRNA, or detailed investigation of a few marker proteins, usually major products of highly specialized differentiated cells. Studies of RNA–DNA hybridization are discussed by Knowler in Chapter 7 of this book. Examples of uterine estrogen-regulated proteins are presented in this chapter, concentrating on estrogen-regulated creatine kinase, originally described by Notides and Gorski (1966) under the name "estrogen-induced protein" (IP). This protein was discovered by pursuing the observations (see review by Katzenellenbogen and Gorski, 1975) that the characteristic early estrogen-stimulated increases in phospholipid and RNA synthesis can be blocked by inhibitors of protein, or of RNA synthesis. The interpretation was that new mRNA, for proteins needed to mediate the early metabolic changes in the uterus, is rapidly synthesized in response to estrogen. Subsequently, Notides and Gorski (1966) demonstrated the rapid stimulation of synthesis of a specific uterine "estrogen-induced protein" (IP), a possible candidate for such a regulator protein. However, no evidence was obtained

Alvin M. Kaye ● Department of Hormone Research, Weizmann Institute of Science, Rehovot 76100, Israel. This review is dedicated to the memory of Prof. Hans R. Lindner, mentor and colleague, whose untimely death on November 19, 1982 has left an irreparable gap in the field of hormone research and in the hearts of his friends around the world.

for this role for IP. Baulieu et al. (1972) postulated that IP might be a "key intermediary protein" needed to stimulate ribosomal RNA synthesis. This hypothesis was rejected since no IP was found in the nucleus (King et al., 1974; Pennequin et al., 1975), the predicted site of action for such a key protein. Phosphoprotein phosphatase activity was found in purified IP preparations (Vokaer et al., 1974), but was later separated from IP (Kaye et al., 1975b). A highly recommended comprehensive review of the literature prior to 1976 on estrogen effects on the uterus is given by Segal et al. (1977).

Estrogen-induced protein became a favorite marker protein for estrogen and antiestrogen action in the rat uterus because of the rapidity (within 40–60 min) with which an increase in its rate of synthesis could be detected after estrogen treatment (Barnea and Gorski, 1970), and its induction in vitro (Mayol and Thayer, 1970; Katzenellenbogen and Gorski, 1972). This extreme speed of induction holds promise for its eventual use in a cell-free system for probing the molecular biology of the interaction of an estrogen–receptor complex with the "acceptor" sites on chromatin, presently the unknown "black box" of steroid hormone action. Hopefully, a cell-free system can be developed which would survive long enough to permit the detection and analysis of such a rapid hormonal response.

2. Sequential Development of Estrogen Responsiveness

The uterus of the rat provides an extremely favorable system for the study of gene action since it shows a progressive and differential gene activation by estrogen during the period from birth to puberty. Responsiveness to a single dose of estrogen can be shown to be acquired in distinct developmental stages when individual components of the overall estrogen response are analyzed (Sömjen et al., 1973b; Kaye et al., 1972, 1974, 1975a, 1980a; Katzenellenbogen and Greger, 1974; Walker et al., 1976, 1978; Peleg et al., 1979).

In perinatal life, although estrogen receptors are present in uterine cytoplasm (Sömjen et al., 1976) and are capable of being transferred into the nucleus (Sömjen et al., 1974), there is a period of approximately 2 weeks after birth during which there is a stimulation of synthesis of a limited number of proteins following a single estrogen injection (Sömjen et al., 1973b). Bulk protein and RNA synthesis can be stimulated by estrogen only after the second week of life. DNA synthesis is accelerated by estrogen only in uteri of rats that are 20 days of age or older (Kaye et al., 1972).

2.1. Perinatal Stage of Responsiveness: Stimulation of Synthesis of a Small Number of Proteins

Two receptors and two enzymes can presently be identified in this first stage of responsiveness to estrogen. The replenishment of cytoplasmic estrogen receptors following estrogen administration is due, at least in part, to synthesis of receptor proteins (Gorski *et al.*, 1971; Cidlowski and Muldoon, 1978). Replenishment takes place at 6 days to essentially the same extent and with the same time course as at 10 and 20 days (Peleg *et al.*, 1979).

Progesterone receptor synthesis is also dependent on estrogen (cf. Leavitt *et al.*, 1977). In 4-day-old rats, there is an approximately sixfold induction of uterine progesterone receptors by estrogen (Raynaud *et al.*, 1980). In fetal guinea pig uterus, Pasqualini and his colleagues (Gulino *et al.*, 1981) demonstrated an increase in progesterone-receptor concentration following estrogen injection into pregnant females.

The enzyme whose induction has been demonstrated at the earliest postnatal age is ornithine decarboxylase, the rate-limiting enzyme in polyamine synthesis. In rats the age of 2 days, the specific activity of ornithine decarboxylase induced by estrogen injection is indistinguishable from the specific activity of ornithine decarboxylase attained after induction at 21 days (Kaye *et al.*, 1973).

The estrogen-induced protein, which can be detected by double isotope labeling (see reviews by Katzenellenbogen and Gorski, 1975; Galand *et al.*, 1978) or by fluorography (Walker *et al.*, 1976, 1979) of ^{35}S-labeled proteins separated by sodium dodecylsulfate (SDS)–polyacrylamide gel electrophoresis (PAGE), is induced as early as 5 days after birth, in Wistar-derived rats (Walker *et al.*, 1976). It was previously reported by Katzenellenbogen and Greger (1974) that IP was induced by estrogen in 6-day-old Sprague–Dawley rats. During the period between 5 and 10 days after birth the inducibility of IP by estrogen increases (Walker *et al.*, 1976; Kaye *et al.*, 1980a, 1981).

2.2. Intermediate Stage of Responsiveness: Stimulation of All Macromolecular Synthesis Except DNA

At 5–10 days after birth, estradiol-17β does not cause a significant increase in the weight, protein, RNA, or DNA content of Wistar rat uteri (Kaye *et al.*, 1974; Sömjen *et al.*, 1973b) when measured 24 hr after administration, the time of maximal effect in older rats. However, at the age of 15 days there is a significant increase in wet weight, including increases in both protein and RNA content, with no change in the content of DNA.

In addition to the change in responsiveness to estrogen shown by IP synthesis during the period between 5 and 10 days after birth (Section 2.1), Katzenellenbogen and Greger (1974) found that estradiol causes a very small increase in 2-deoxyglucose phosphorylation at the age of 9 days. This response is capable of maximal stimulation at 19 days.

An example of an estrogen-regulated enzyme which shows a response to estrogen from the age of 11 days only is uterine peroxidase (Lyttle *et al.*, 1979), an enzyme which has been studied extensively as a marker for estrogen responsiveness (cf. Jellinck *et al.*, 1979; Lyttle and De Sombre, 1977).

2.3. The Stage of Complete Responsiveness Including DNA Synthesis

The presumptive replicative polymerase, DNA polymerase α (Weissbach, 1977) shows an increase in activity in immature rat uterus by 16 hr following estrogen administration (Harris and Gorski, 1978a, 1978b; Walker et al., 1978). The time course of stimulation of DNA polymerase α parallels that of thymidine incorporation into DNA following estradiol stimulation (Kaye et al., 1972). Moreover, although uterine DNA polymerase activity can be stimulated by doses of estrogen as low as 0.6 ng/g body weight in 20- and 25-day-old rats, doses as high as 170 ng/g body weight failed to cause any increase in DNA polymerase α activity in uteri of 10- or 15-day-old rats. Thus, the ability to respond to estrogen by both an increase in thymidine incorporation into DNA and an increase in DNA polymerase α activity develop in parallel between 15 and 20 days after birth in Wistar-derived rats (cf. review of Stormshak et al., 1978).

The acquisition of responsiveness to estrogen during development is not restricted to the rat uterus. In fetal guinea pig, Pasqualini and Sumida and their colleagues have reported an increased concentration of progesterone receptors unaccompanied by an increase in DNA content, after estrogen injection (Gulino et al., 1981). They compare this situation to the data on the rat, discussed above, and conclude that "responsiveness of the uterus to estrogen stimulation may be a heterogeneous process that begins to appear already during the fetal period and that develops progressively."

2.4. "Early" and "Late" Responses to Estrogen

In another time frame, the responses to a single injection of estradiol by a fully competent rat (20 days or older) appear at varying times after the stimulus and have often been divided into "early" and "late" responses. There is a rough parallelism between the sequence of these reactions and the sequential development of responsiveness during postnatal ontogeny. The induced protein, the first macromolecular synthetic marker of estrogen action, appears within 40–60 min (Barnea and Gorski, 1970; Mayol and Thayer, 1970). Evidence from inhibitor studies for the production of mRNA for IP synthesis within 15 min of estrogen injection has been presented (De Angelo and Gorski, 1970; Katzenellenbogen and Gorski, 1972). An increase in the synthetic rate for general protein synthesis (see review by Katzenellenbogen and Gorski, 1975) begins from 2 to 4 hr after estrogen treatment and continues for several hours. DNA synthesis peaks 24 hr after estrogen injection (Kaye et al., 1972). Some of the enzymes whose time course of increase in activity after estrogen stimulation has been measured are listed in Table I, in the approximate order of their appearance.

Table I. Examples of Uterine Enzyme Activities Stimulated by a Single Dose of Estradiol-17β[a]

Enzyme	Test system	Dosage	Earliest response (hr)	Earlier(est) time tested (hr)	Peak response (hr)	Reference
t-boc-Ala-Ala-Pro-Ala AMC hydrolase	Weanling mice	0.15 μg/mouse	0.5	0.5	6	Katz et al., 1980
Ornithine decarboxylase	20-day-old rats	0.5 μg/rat	2	1	4	Kaye et al., 1971
	ovex or hypox[a]	5 μg/rat	4	4	—	Cohen et al., 1970
SAM decarboxylase	20-day-old rats	0.5 μg/rat	3	1	5	Kaye et al., 1971
Mg^{2+}-dependent ATPase	Ovex	Not given	3–6	1–2	—	Tam and Spaziani, 1970
Aldolase	Ovex	10 μg/100 g rat	4	2	16	Schwark et al., 1969
Hexokinase	Ovex	10 μg/100 g rat	4	2	16	Valdares et al., 1968
Poly(adenosine diphosphoribose synthetase)	Immature mice	2 μg/mouse	6	3	18	Miura et al., 1972
Pyruvate kinase	Ovex	15 μg/100 g rat	8	4	16	de Asua et al., 1968
Aspartate transcarbamylase	20–22-day-old rats	10 μg/rat	12	12	—	Tremblay and Thayer, 1964
Leucine aminopeptidase	Ovex	1 μg/rat	12	6	24	Schmidt et al., 1967
Isocitrate dehydrogenase	Ovex	1 μg/rat	12	6	48	Schmidt et al., 1967
Glucose-6-PO_4 dehydrogenase	Ovex	1 μg/rat	12	6	24	Schmidt et al., 1967
DNA polymerase α	20-day-old rats	5 μg/rat	16	12	20–36	Walker et al., 1978
	21-day-old rats	1 μg/rat	18	12	24	Harris and Gorski, 1978
Lysyl oxidase	Mouse (cervix)	1 μg/mouse	12–18	12	18	Ozasa et al., 1981
Succinic dehydrogenase	Ovex	30 μg/rat	24	24	48	Eckstein and Villee, 1966

[a] Ovex, ovariectomized adult rats; hypox, hypophysectomized adult rats.

3. Estrogen-Responsive Creatine Kinase

The estrogen-"induced protein" (IP) from rat uterus was originally characterized as a single polypeptide chain with a mol. wt. of approximately 46,000 and an isoelectric point of approximately 4.5 (for short reviews see Kaye *et al.*, 1975a; Galand *et al.*, 1978). Estrogen-induced protein synthesis is stimulated by estrogen in both the endometrium and myometrium of 22- to 25-day-old rats (Katzenellenbogen and Leake, 1974) and in the epithelium, stroma, and myometrium of mature ovariectomized rats (Dupont-Mairesse and Galand, 1975; Mairesee and Galand, 1982).

3.1. Techniques Permitting Localization of IP to a Single Protein Band

Originally, Notides and Gorski (1966) used the isotope ratio method and starch gel electrophoresis to reveal an IP region migrating more rapidly than albumin. Similarly, polyacrylamide gels, run under nondenaturing conditions were employed. The use of gelatinized cellulose acetate, (Cellogel, Chemetron, Milan, Italy) in the form of 250-μm-thick analytical strips (Sömjen *et al.*, 1973a) and as 0.25- and 0.5-cm-thick preparative blocks (King *et al.*, 1974), as the support medium for electrophoresis, was found to be quick, convenient, and sensitive. The ability simply to squeeze the IP from the appropriate region of the gel (10% faster migration than a bovine serum albumin marker) resulted in an order of magnitude purification (Walker *et al.*, 1979; Kaye *et al.*, 1980a) as judged by analysis of [^{35}S]methionine-labeled IP fractions by SDS–PAGE.

The switch from double-labeling to [^{35}S]methionine labeling was based on the following considerations. The use of double-isotope labeling for studying the induced protein had the advantage of sensitivity and reproducibility. However, since it is a ratio method, quantitation of the rate of synthesis is only approximate and no information is available for calculating the actual concentration of IP either in the control, unstimulated uterus, or in the estrogen-stimulated uterus. The use of [^{35}S]methionine incorporation combined with SDS–PAGE permitted the identification of a major protein band, migrating between tubulin and actin, as the band that contained IP (Walker *et al.*, 1976). Photometric evaluation of the fluorograms of SDS–PAGE gels permitted the estimation of the ^{35}S incorporation into the IP band as a proportion of the total incorporation into newly synthesized proteins. This technique led to the identification of a protein-staining band (46,000 daltons) corresponding to the radioautographically detectable band. Moreover, this band was present in control uteri of immature rats, unstimulated by exogenous estrogen (Walker *et al.*, 1976). The rate of [^{35}S]methionine incorporation into uterine proteins of such control rats was approximately half of that in uteri of estrogen-stimulated rats, indicating that there is a high rate of constitutive synthesis of IP in the uterus (see Section 3.5 for more precise quantitation).

3.2. "Estrogen-Induced Protein" is Not Confined to the Rat Uterus

The finding of constitutive synthesis of IP suggested (Walker *et al.*, 1976) that the function of IP may not be limited to the estrogen-stimulated state, nor to the uterus and the vagina (Katzman et al., 1971).

A survey of several organs for the constitutive presence of IP (Fig. 1) led to its detection in pituitary gland, as well as in the hypothalamus, in the cerebral cortex of both male and female rats, and in the liver and muscle in much lower concentrations (Walker *et al.*, 1979). Therefore, the protein in rat hypothalamus with electrophoretic behavior identical to the uterine IP (Walker *et al.*, 1976) was investigated. The uterine and brain IPs were found to be indistinguishable (Walker *et al.*, 1979) by sequential separation according to charge (Cellogel) and size (SDS–PAGE) and upon comparison of limited protease digestion patterns

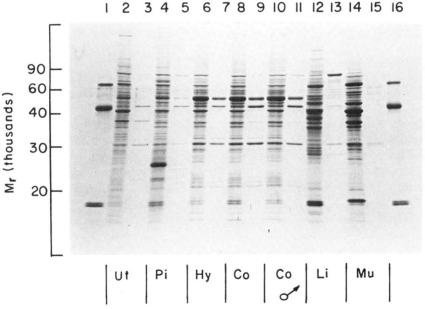

Figure 1. Coomassie brilliant blue staining pattern of SDS–polyacrylamide gel electropherograms of unfractionated cytosol (even lanes 2–14) and cellogel fractionated cytosol (odd lanes 3–15) of uterus (Ut), pituitary (Pi), hypothalamus (Hy), cerebral cortex (Co), cerebral cortex of male (Co ♂), liver (Li), and muscle (Mu). Lanes 1 and 16 show mol. wt. markers (BSA, 67,000; ovalbumin, 45,000; hemoglobin, 16,000). IP migrates slightly more slowly than ovalbumin. (From Walker *et al.*, 1979.)

following incubation with *Staphylococcus aureus* V8 protease (Cleveland *et al.*, 1977).

The high-speed supernatant solution of whole brains of 25-day-old rats was chosen as the source of IP for purification (Kaye and Reiss, 1980) by ammonium sulfate precipitation, diethylaminoethyl (DEAE) cellulose chromatography, and preparative SDS–PAGE. Antibodies were raised in rabbits against this rat brain IP. The antiserum was shown to precipitate IP specifically from ^{35}S labeled uterine cytosols, providing additional (immunochemical) data to substantiate the conclusion that uterine and brain IP are identical.

3.3. Enolase γγ ("Neuron-Specific Enolase," Antigen 14-3-2) is a Component of IP

Since IP was found to be one of the major soluble proteins in the rat brain, the neurochemical literature was searched for a description of a brain protein with physicochemical characteristics similar to IP.

In 1968, two groups described the isolation of acidic proteins from rat (Bennett and Edelman, 1968) and bovine brain (Moore and Perez, 1968) which were called antigen α and 14-3-2, respectively. There proteins have very similar amino acid compositions (Marangos *et al.*, 1975b) and electrophoretic mobilites, (Bennett, 1974) and show high immunological cross-reactivity; therefore antigen α and 14-3-2 are thought to be identical. Antigen α was originally reported to be brain-specific (Bennett and Edelman, 1968; Cicero *et al.*, 1970b) but not species-specific. The presence of 14-3-2 in the brain of cat (Marangos *et al.*, 1975a), chicken (Cicero *et al.*, 1970a), and other species has been reported.

This wide distribution and apparent specificity for the brain has been explained by the observation that 14-3-2 preparations show enolase (2-phospho-D-glycerate-hydrolase, EC 4.2.1.11) activity (Bock and Dissing, 1975) under conditions of purification which eliminated detectable contaminants (Marangos *et al.*, 1976). The rat brain contains three enolase isozymes of which the most acidic (γγ dimer) was reported to be brain-specific (Marangos *et al.*, 1976; Bock *et al.*, 1978). The other isozymes are a neutral form (αα dimer) which is not brain-specific, and an intermediate form, apparently a hybrid dimer (αγ).

The isoelectric point of 14-3-2 is 4.7 (Marangos *et al.*, 1975b) and its mol. wt. is 48,000–50,000 (Moore and Perez, 1968; Marangos *et al.*, 1975b). The concentration of 14-3-2 in rat brain was found to be approximately 1% of the soluble brain proteins (Bennett, 1974; Marangos *et al.*, 1975a).

With all of these indications of a similarity between IP and the acidic isozyme of enolase, a comparison of the enolase isozyme composition of brain and uterine cytosol was made (Kaye *et al.*, 1980b). The uterus, as well as the brain, was found to contain all three isozymes of enolase, αα, αγ, and γγ. In the uterus, the αγ peak is smaller than in the brain; the reproducibly detectable γγ peak accounts for <1% of the total enolase activity.

The responsiveness of enolase isozymes in uterus and brain of 30-day-old rats was tested during the first 24 hr after injection of estradiol-17β. While no significant changes were observed in uterine neutral and intermediate enolases, the acidic (γγ) isozyme as a percentage of the total enolase showed a 40% increase at 2 hr, 150% at 12 h, and 170% at 24 hr after estradiol injection.

A component of IP was therefore identified as enolase γγ (Kaye *et al.*, 1980b; Reiss and Kaye, 1981) on the basis of:

1. The specific enolase activity of the purified product, which is higher than reported previously (Marangos *et al.*, 1976, 1978).
2. The position of elution from a DEAE cellulose column which corresponds to that of the γγ isozyme of enolase.
3. The increased proportion of total uterine enolase activity represented by enolase γγ after estradiol treatment.
4. The increase in the double isotope ratio of enolase γγ in uterine cytosol obtained 1 hr after injection of estradiol.
5. The precipitation by antibodies against purified γγ enolase supplied by Dr. Paul Marangos (N.I.M.H., Bethesda).

3.4. The Major Component of IP is the Brain Type (BB) Isozyme of Creatine Kinase

The small proportion of enolase activity in the uterus represented by the γγ form, as well as the demonstration by two-dimensional gel separations of uterine cytosol that IP consisted of more than a single protein (Kaye *et al.*,

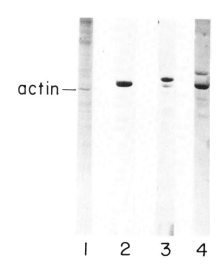

Figure 2. Resolution of IP into creatine kinase B and enolase γγ bands by modified SDS–polyacrylamide gel electrophoresis. Samples of IP (3 μg) from a DEAE cellulose chromatography step of purification were analyzed by SDS–PAGE (lanes 2–4) along with uterine cytosol to provide a reference (actin) band (lane 1). Lanes 1 and 2, 10–20% polyacrylamide gradient gel under the conditions described by Laemmli (1971). Lane 3, gel conditions as lanes 1 and 2 except that the concentration of Tris buffers in the stacking and resolving gels and the Tris–glycine buffer in the electrode chambers were all reduced in concentration by half. Lane 4, gel conditions as in lanes 1 and 2, except that concentrations of all buffer systems and the bisacrylamide were increased (Reiss and Kaye, 1981). The gels were stained with Coomassie brilliant blue. (From Reiss and Kaye, 1981.)

1980a) led to attempts at further purification of IP and to a resolution of enolase activity from a presumably major component. Changes in conditions for SDS–PAGE indeed permitted separation of brain IP into two components (Fig. 2), one of which was present in approximately four to five times the concentration of the other (Reiss and Kaye, 1981). The adsorption of IP on reactive blue 2 agarose permitted the preparative separation of the two components (Fig. 3) and the direct confirmation that the minor component was enolase (see Section 3.3.).

During the search for improved purification methods for enolase, we saw the report of Wood (1963, 1964) that enolase was found to accompany creatine kinase (ATP: creatine phosphotransferase, EC 2.7.3.2) up to the last stage of its purification. By testing the major component of IP for creatine kinase activity, we showed that it was indeed the BB isozyme of creatine kinase (CK-BB). The increase in the activity of uterine CK–BB following estrogen injection could be followed simply by activity assays of total uterine cytosol, since in both uterine and brain cytosols, essentially all of the CK activity is in the BB form (Reiss and Kaye, 1981).

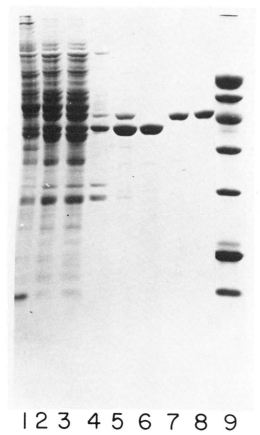

Figure 3. Visualization of steps in the simultaneous purification of creatine kinase BB and enolase γγ from rat brain. (Reiss and Kaye, 1981.) Samples from successive steps were subjected to SDS polyacrylamide gel electrophoresis using a concentrated buffer system which separates creatine kinase B from enolase γ subunits. The gel was stained with Coomassie brilliant blue. Lane 1: 20 μg of 27,000g supernatant fraction of brain homogenate; lane 2: 20 μg of the material precipitated between 40 and 65% saturation of ammonium sulfate; lane 3: 20 μg of supernatant after pH 5.2 precipitation; lane 4: 10 μg of the IP fraction from step gradient elution of a DEAE cellulose column; lane 5: 10 μg of the pooled IP fraction after linear salt gradient elution from a DEAE cellulose column; lane 6: 4 μg of creatine kinase BB eluted from a Reactive Blue-2 agarose column; lane 7: 2 μg of enolase γγ not adsorbed to the Reactive Blue-2 agarose column; lane 8: 2 μg of enolase γγ after additional DEAE cellulose linear salt gradient chromatography; lane 9: mol. wt. markers: BSA, 67,000; catalase, 60,000; ovalbumin, 45,000; glyceraldehyde-3-phosphate dehydrogenase, 36,000; carbonic anhydrase, 30,000; myoglobulin, 17,200; cytochrome c, 12,000. (From Reiss and Kaye, 1981.)

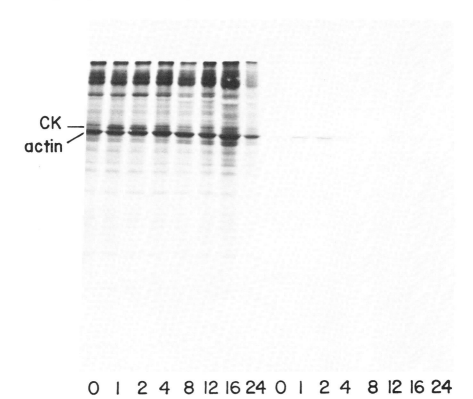

CK

actin

0 1 2 4 8 12 16 24 0 1 2 4 8 12 16 24
Time after E$_2$ injection (h)

Figure 4. Fluorographic analysis of uterine creatine kinase response to estradiol-17β using immunoprecipitation with anticreatine kinase BB antisera. Rats received intraperitoneally 5 μg estradiol-17β (except for control rats) and were killed at the specified times. Uteri were labeled with [^{35}S]methionine (Kaye *et al.*, 1980a). The cytosols (9 × 10^5 cpm) were subjected to an unrelated immune reaction (Caravatti *et al.*, 1979) using 100 ng BSA and 1 μl of anti-BSA antisera. The unrelated immune complexes were removed using 10 μl of 10% formalin-inactivated *S. aureus* (Kessler, 1975). The supernatant solutions were treated for 60 min at 37°C with 2 μl of anticreatine kinase BB and then incubated overnight at 4°C. The immune complexes were precipitated using 20 μl of 10% *S. aureus* and applied to SDS–PAGE using a concentrated buffer system (Reiss and Kaye, 1981). The first eight lanes (left) show cytosol proteins before immunoprecipitation. The last eight lanes show the corresponding immunoprecipitates of creatine kinase B subunits. The presence of the large excess of immunoglobulin heavy chain (50,000 M$_r$) caused a displacement of the creatine kinase band which appears to migrate at a lower molecular weight. CK, creatine kinase BB; E$_2$, estradiol-17β. (From Reiss and Kaye, 1981.)

The major component of IP was identified as creatine kinase BB on the basis of:

1. Its specific activity which is among the highest reported for purified creatine kinase BB (Eppenberger *et al.*, 1967; Focant and Watts, 1973).

2. The identical position of elution of uterine and brain creatine kinase from a DEAE cellulose column, which coincides with the peak of increased double-isotope ratio characteristic of IP.

3. The limited protease digestion pattern of the purified enzyme which is indistinguishable from the pattern obtained in previous analyses of uterine IP (Walker et al., 1979).

4. The increase in specific activity of creatine kinase BB in the uterus after injection of estradiol or stimulation of estrogen secretion.

5. Specific immunoprecipitation of uterine IP with antibodies against native brain creatine kinase (Fig. 4).

The greater intensity of the precipitated creatine kinase at 1–2 hr after estrogen treatment and its rapid decay, starting earlier than 4 hr after estrogen injection, was in complete accord with the evidence for a rapid but transient increase in the rate of IP synthesis obtained by other methods (Barnea and Gorski, 1970; Walker et al., 1976).

3.5. Synthesis of mRNA for Creatine Kinase BB is Stimulated by Estrogen

The problem of extracting active mRNA from rat uterus was solved by using the phenol–cresol method of Kirby (1968) with some modifications (Walker and Kaye, 1981). The Poly(A)-rich RNA fraction obtained by oligo (dT)-cellulose chromatography (Aviv and Leder, 1972) was translated in the micrococcal nuclease-treated (Pelham and Jackson, 1976) rabbit reticulocyte cell-free protein-synthesizing system and immunoprecipitated with antiserum directed against IP. A band with mobility in SDS–PAGE identical to IP was found (Fig. 5). In addition, two-dimensional gel analysis of [^{35}S]-labeled proteins synthesized by surviving uteri (Fig. 6) or by the reticulocyte lysate system in response to uterine mRNA (Fig. 7), was used to identify and isolate the newly synthesized creatine kinase BB. Measurement of the incorporation of [^{35}S]methionine into the CK spot showed that the extent of induction of this protein, 1 hr after estrogen injection was fourfold in the surviving uterus and 2.3-fold in cell-free translation products. Comparison of the limited protease digestion patterns of CK synthesized in vivo and in the cell-free translation system (Fig. 8) with creatine kinase isolated from rat brain (Reiss and Kaye, 1981) confirmed the identity of the product of synthesis in the reticulocyte lysate system as CK–BB. Thus, the concentration of translatable mRNA for CK is specifically elevated in the uterus, 1 hr after estrogen administration. If "induction" of protein synthesis is defined as the increase in concentration of a protein by means of an increased concentration of translatable mRNA for that protein, then the experiments described in this section justify, in retrospect, the terms "induced protein" for creatine kinase BB in the rat uterus. These two terms can now be used synonymously, since in the rat uterus the

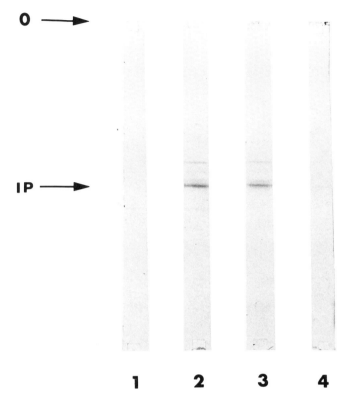

Figure 5. Fluorogram of SDS–polyacrylamide gel electrophoresis of immunoprecipitated products of cell-free translation of brain and uterine mRNA. mRNA was prepared from brain and uterus of 25-day-old rats by phenol–cresol extraction and oligo (dT) cellulose chromatography. The mRNA was incubated in a rabbit reticulocyte cell-free protein synthesizing system (Pelham and Jackson, 1976) in the presence of [^{35}S]methionine. Translation products were treated with antiserum directed against IP or an unrelated immune serum (anti IgA). Immune precipitates were subjected to electrophoresis and the [^{35}S]methionine-labeled proteins were visualized by fluorography. Lanes 1–3: translation products from incubations with no added mRNA (lane 1), brain mRNA (lane 2), and uterine mRNA (lane 3) were precipitated by anti-IP serum. Lane 4 shows translation products of uterine mRNA precipitated by anti-IgA serum. The starting point of migration and the mobility of IP are indicated by arrows. (From Walker and Kaye, 1981.)

minor component of IP, enolase γγ, accounts for only a fraction of a percent of IP.

In terms of understanding gene regulation, it may be that the case of CK in the uterus, which is regulated between a high constitutive level and an even higher induced level, represents a more common situation than an all-or-none action of "switching on" or "switching off" a gene. This system, therefore, may provide new insights into gene regulation compared to the well-studied systems

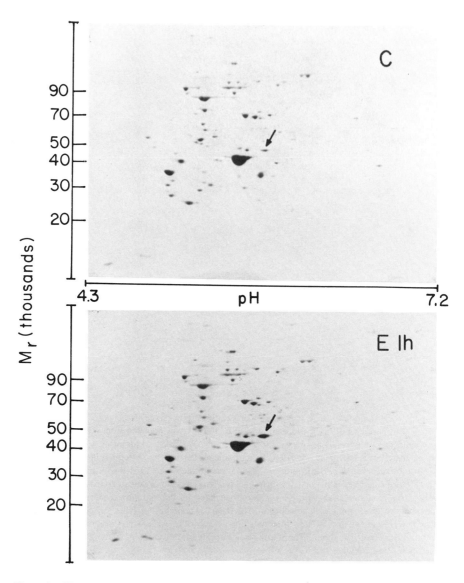

Figure 6.   Fluorogram of two-dimensional polyacrylamide gel separation of uterine cytosol proteins. Uteri of untreated rats (C) or rats treated 1 hr previously with 5 μg of estradiol-17β (E) were incubated for 2 hr in the presence of ^{35}S-labeled methionine (Walker *et al.*, 1976). Cytosol was prepared following the incubation and samples containing equal amounts of radioactivity were subjected to two-dimensional separation according to O'Farrell (1975). The first dimension was isoelectric focusing and the second was electrophoresis in the presence of SDS. The origin of migration is at upper right of the figure; the arrows indicate the position of IP. (From Walker and Kaye, 1981.)

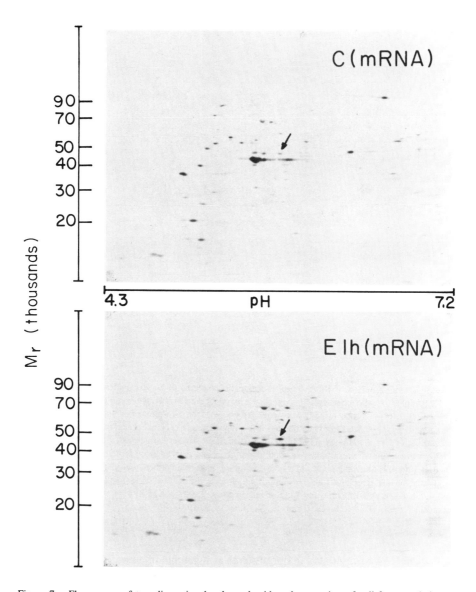

Figure 7. Fluorogram of two-dimensional polyacrylamide gel separation of cell-free translation products directed by uterine mRNA. Uteri were excised from 25-day-old rats, 1 hr after injection with 1% ethanol vehicle (C) or 5 μg of estradiol-17β (E), and frozen immediately in liquid nitrogen. mRNA was prepared by phenol–cresol extraction and oligo (dT) cellulose chromatography and added to a rabbit reticulocyte cell-free system (Pelham and Jackson, 1976). Reaction mixtures were subjected to two-dimensional gel separation according to O'Farrell (1975) as described in the legend to Fig. 6. The arrows indicate the position of IP. (From Walker and Kaye, 1981.)

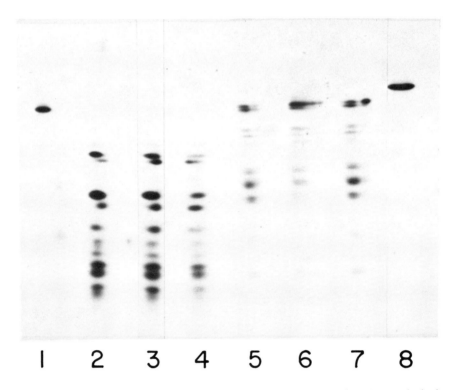

Figure 8.   Fluorogram of limited protease digestion patterns of *in vivo* and *in vitro*–synthesized CK. Samples of radioactive protein synthesized by intact uteri or directed by uterine or brain mRNA in the reticulocyte lysate system were coelectrophoresed on two-dimensional gels with unlabeled brain proteins. Spots were cut from the gels following staining and partial protease digestion performed according to Cleveland *et al.* (1977) on a 15–20% SDS–polyacrylamide gel. Lanes 2–4: digestion pattern of proteins corresponding in mobility to actin and derived from intact uteri, brain mRNA and uterine mRNA respectively; lanes 5–7: digestion pattern of proteins corresponding in mobility to IP and derived from intact uteri, brain mRNA, and uterine mRNA; lanes 1 and 8: protein markers: undigested samples of actin and tubulin respectively. Lanes 3 and 8 were subjected to shorter exposures than the other lanes. (From Walker and Kaye, 1981.)

of such estrogen-regulated major proteins as the avian or amphibian egg proteins (see reviews by Chan and O'Malley, 1976; Higgins and Gehring, 1978).

3.6.   Distribution of Estrogen-Regulated CK and Putative Related Proteins

Regulation of uterine CK activity by estrogen is not confined to the rat. In several mouse strains, increased synthesis of IP has been demonstrated following injection of estradiol-17β (Kaye *et al.*, 1975a) but with double-labeling peaks only approximately one-fifth of the height of the peak obtainable in the rat. Mouse IP has also been demonstrated immunologically (Iacobelli *et al.*, 1979) and fluorographically (Kaye and Reiss, 1980). The species difference may now be explained by the finding that the constitutive creatine kinase activity in mouse uterus is higher than in the rat, making the increase in CK activity to a level slightly higher than the stimulated level in the rat, a smaller proportional rise (A. Shaer and A.M. Kaye, unpublished observations). Most recently, a rise in CK activity has been reported in human endometrium during the late luteal phase of the menstrual cycle (Kaye *et al.*, 1981).

Long-term effects of sex steroid hormones on creatine kinase activity, in species other than the rat, are the two- to threefold increase in CK–BB activity found in bovine myometrium during early pregnancy (90–120 days), by Iyengar and Iyengar (1980) and Iyengar *et al.* (1980) and the doubling of human myometrial CK during late pregnancy, reported as early as 1965 by Luh and Henkel.

The characterization of IP by sequential separation based on charge and size (Walker *et al.*, 1979; Kaye *et al.*, 1980a; Kaye and Reiss, 1980; Korach *et al.*, 1981) or on partial purification followed by two-dimensional gel analysis (Skipper *et al.*, 1980) permits the testing of apparent "induced proteins" such as electrophoretic peaks reported in the brain (Thomas and Knight, 1978), rat mammary gland (Mairesse *et al.*, 1977), or human endometrium (Iacobelli *et al.*, 1981) for their relationship to creatine kinase and/or enolase. Very recently, Vertes et al. (1981) have reported that, by the double-labeling technique and SDS–PAGE, or electrophoresis on cellogel, induction of IP was found in rat anterior hypothalamus–preoptic area and median eminence. It was therefore of great interest to determine whether CK or enolase or both are induced by estrogen in specific brain regions where a substantial part of the protein identified electrophoretically as IP is enolase γγ.

Recent experiments using PAGE techniques which resolve CKBB and enolase γγ (Malnick, Soreq and Kaye, in preparation) demonstrate an increased rate of synthesis of CKBB but not of enolase γγ in hypothalamic regions of immature rat brains, one hour after estrogen injection. Indeed, the recent findings that the rate of synthesis of CKBB can be stimulated by estrogen in immature rat ovaries (Malnick and Kaye, 1982) and that CKBB activity can be increased in fragments of normal human breast tissue upon incubation *in vitro* in 3×10^{-8} M estradiol 17-β (Shaer *et al.*, 1982) make it likely that all the organs of the female reproductive system can respond to estrogen stimulation by increased synthesis of CKBB.

In the rat uterus, a protein whose rate of synthesis was stimulated by the so-called "antiestrogen" nafoxidine, (Mairesse and Galand, 1979) was shown to be resolved into two proteins, of 27,000 and 30,000 apparent mol. wt., by SDS–PAGE of proteins recovered from electrophoresis on cellogel (Mairesse *et al.*, 1981). These two proteins were not precipitated by antibodies against IP, were constitutively present in uterus, and, like CK, were constitutively present in brain in higher concentrations than in uterus. This finding, that an "antiestrogen" or "impeded estrogen" stimulates the synthesis of proteins differently from estrogen, may provide a new approach to the study of gene activation in the uterus.

3.7. Possible Function of Estrogen-Regulated CK in the Uterus

Creatine kinase regulates the intracellular concentration of ATP by what has been termed a "buffering" action (Jacobus and Lehninger, 1973). Excess ATP formed during glycolysis or fatty acid catabolism is converted to creatine phosphate which, in turn, is used for rapidly regenerating ATP. In the uterus, the increased glucose uptake and metabolism during the early stage of response to estrogen (Smith and Gorski, 1968) may provide an excess of energy supply, stored in the form of creatine phosphate, which subsequently may be used in the processes of growth and cell division. On the other hand, a very high energy utilization by uterine cells during the "early" response to estrogen may be accompanied by a reduction of the energy stored as creatine phosphate.

To determine if the rapid change in synthesis of CK–BB in the rat uterus

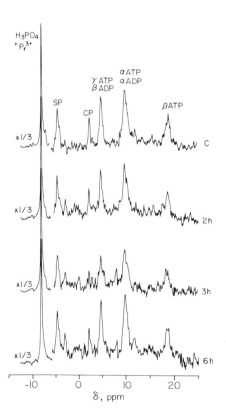

Figure 9. ^{31}P NMR spectra of soluble uterine phosphates. Immature female rats were injected i.p. with 5 μg of estradiol-17β and killed at the indicated times. Uteri (12) were immediately immersed in Krebs–Hepes buffer (pH 7.4), containing 20% D_2O, in 10-mm NMR tubes with a coaxial external reference (1-mm diameter tube containing 1 M $H_3 PO_4$ and 0.11 M $PrCl_3$ represented at 1/3 of the intensity of the tissue spectra) and oxygenated with 95% O_2, 5% CO_2. Spectra were recorded at 4°C from a Bruker WH270 pulse-FT spectrometer at 109.3 mHz. 3000 scans were accumulated at a tip angle of 42° and an acquisition time of 0.68 sec. C, control; CP, creatine phosphate; SP, sugar phosphates.

after estrogen injection has a significant physiological role in the early uterotropic action of estrogen, changes in the concentrations of the high-energy phosphate compounds in the uterus, creatine phosphate, ATP, and ADP, were followed by ^{31}P nuclear magnetic resonance spectroscopy (NMR) as a function of time after estrogen administration. Within the first 1.5–3 hr after injection of estradiol-17β into 25- to 27-day-old rats, the concentrations of CP, ATP, and ADP were found to decline by approximately 40% (Fig. 9) followed by a return to control values at about 12 hr (H. Degani, A. Shaer, T. Victor and A.M. Kaye, in preparation). The same early decrease in creatine phosphate concentration was observed in adult ovariectomized rat uteri. The use of NMR to study changes in creatine phosphate and other phosphorus-containing compounds both *in vivo* and in surviving uteri offers an opportunity to study the physiological consequences of the extremely rapid changes in expression shown by the estrogen-responsive creatine kinase gene.

4. Regulation of Glucose-6-Phosphate Dehydrogenase

In the case of glucose-6-phosphate dehydrogenase (G6PD), the ability of estrogen to increase its activity in the uterus of ovariectomized adult rats has long been known (Scott and Lisi, 1960; Barker and Warren, 1966); the detailed mechanism of its induction is now emerging. This enzyme is the first in the pentose phosphate pathway of glucose utilization which leads to the production of ribose as well as NADPH (see review by Levy, 1979). The following resume of recent results showing that G6PD stimulation is a consequence of both an increased production of mRNA and of posttranslational events is based mainly on the account of Barker *et al.*, (1981). G6PD is presented as the other example of an estrogen-induced protein for which the mRNA has been translated in a cell-free system and been shown to increase following estrogen injection.

Previous to this demonstration it had been shown that the estrogen-induced increase in G6PD activity was inhibited by prior treatment with actinomycin D. Estradiol was also shown to inhibit completely the degradation of G6PD (Smith and Barker, 1974, 1977).

To determine the rate of G6PD synthesis, the enzyme was purified from rat liver after induction by a diet "free" of lipids. Antisera against the purified enzyme were made in rabbits and used to measure the rate of incorporation of [^{14}C]amino acids into immunoprecipitable G6PD. Uterine mRNA was isolated by a modification of the guanidine hydrochloride method of Strohman *et al.* (1977), and translated in a messenger-dependent rabbit reticulocyte cell-free protein-synthesizing system (Adams and Barker, 1979). During the first 4 hr after injection of 10 μg estradiol-17β into mature ovariectomized rats, there was no change in the rate of G6PD synthesis, or its total activity. At 6 hr after estrogen treatment there was a fivefold increase in G6PD messenger activity;

this was accompanied by an approximately threefold increase in the rate of G6PD synthesis, which reached a peak value of 18-fold at 12 hr when the increase in mRNA activity was sevenfold. Subsequently, these two parameters declined while the accumulation of total G6PD activity continued, to reach a peak value 36 hr after estrogen injection.

In addition to transcriptional control and regulation of G6PD concentration by prevention of degradation, estradiol was found (Donahue *et al.*, 1979) to decrease the time needed to translate uterine G6PD (from 4.9 min in control, to 2.0 min in 12-hr estrogen-treated rats), and to process G6PD, as measured by the time of first appearance of its characteristic pyroglutamate amino terminal residue (Donohue *et al.*, 1977) following injection of [^{3}H]glutamate (Barker *et al.*, 1981).

5. Considerations for Future Study

The rationale behind the developmental approach to gene regulation, by estrogen, in the uterus is the utilization of a natural test system in which particular responses can be obtained in the absence of the many other processes concerned with growth and cell division that estrogen stimulates. The differentiation of "early" and "late" responses also provides such a system now that it seems clear that separate periods of stimulation are required for "early" responses and the stimulation of DNA synthesis and cell division (Harris and Gorski, 1978b). Other means of dissociating early responses and hyperplastic responses from later events concerned directly with cell replication include the experimental induction of diabetes (Kirkland *et al.*, 1981a) or hypothyroidism (Kirkland *et al.*, 1981b) or by means of hypophysectomy (Sonnenschein and Soto, 1978), or the use of so-called "antiestrogens" (Katzenellenbogen *et al.*, 1979).

The time differential in the first appearance of increased mRNA for the marker proteins, CK and G6PD, provides challenging data for hypotheses and testing. Does the induction of G6PD require previous estrogen stimulation of another protein? Are the regulatory proteins perhaps specific nonhistone proteins (Teng and Hamilton, 1970; Barker, 1971; Kaye *et al.*, 1974; King *et al.*, 1974)? Is there some way in which a group or groups of genes stimulated by estrogen (Table I) are functionally related, as, for example, part of chromatin "domains" (Lawson *et al.*, 1981)?

Both CK and G6PD are processed after translation; G6PD loses its presumed N terminal methionine to acquire a pyroglutamate end (Donahue *et al.*, 1977); CK shows no methionine in the first 40 N-terminal residues when studied by microsequence analysis using [^{3}H]methionine incorporation, a technique that reveals [^{3}H]leucine residues at positions 11 and 22 (M. Walker, N. Reiss, Y. Burstein and A.M. Kaye, unpublished results). Is this the first hint of a relationship between estrogen-induced proteins besides the obvious physicochemical relationship between CK–BB and enolase γγ (Section 3)?

The extremely wide background of experience in the use of the rodent uterus as a test system for investigating hormone action, so briefly touched upon in this chapter, combined with the recent advances described above, makes it possible to presume that the stimulation of gene action by estrogen in this system, envisioned by Mueller *et al.* as early as 1958, will be studied intensively in the future, and that the secrets of gene expression cleverly coiled in the chromosomes of uterine cells will eventually yield to the growing power of molecular biology.

ACKNOWLEDGMENTS. I thank all my collaborators, whose names are included in the references to our joint papers, for their contribution to the work in our laboratory which has been supported in part by grants (to H. R. Lindner) from the Ford Foundation, The Population Council, and The Rockefeller Foundation, New York. A.M.K. is the incumbent of the Joseph Moss Professorial Chair in Molecular Endocrinology.

REFERENCES

Adams, D. J., and Barker, K. L., 1979, Regulation of glucose-6-phosphate dehydrogenase mRNA levels in the uterus by estradiol, *Fed. Proc. Am. Soc. Exp. Biol.* **38**:399.

Aviv, H., and Leder, P., 1972, Purification of biologically active globin messenger RNA by chromatography on oligo-thymidylic acid cellulose, *Proc. Natl. Acad. Sci. USA* **69**:1408–1412.

Barker, K. L., 1971, Estrogen induced synthesis of histones and a specific nonhistone protein in the uterus, *Biochemistry* **10**:284–291.

Barker, K. L., and Warren, J. C., 1966, Estrogen control of uterine carbohydrate metabolism: Factors affecting glucose utilization, *Endocrinology* **79**:1069–1074.

Barker, K. L., Adams, D. J., and Donohue, T. M. Jr., 1981, Regulation of the levels of mRNA for glucose-6-phosphate dehydrogenase and its rate of translation in the uterus by estradiol, in: *Cellular and Molecular Aspects of Implantation* (S. R. Glasser and D. W. Bullock, eds.), pp. 269–281, Plenum Press, New York.

Barnea, A., and Gorski, J., 1970, Estrogen-induced protein. Time course of synthesis, *Biochemistry* **9**:1899–1904.

Baulieu, E. E., Alberga, A., Raynaud-Jammet, C., and Wira, C. R, 1972, New look at the very early steps of oestrogen action in uterus, *Nature New Biol.* **236**:236–239.

Bennett, G. S., 1974, Immunologic and electrophoretic identity between nervous system-specific proteins antigen alpha and 14-3-2, *Brain Res.* **68**:365–369.

Bennett, G. S., and Edelman, G. M., 1968, Isolation of an acidic protein from rat brain, *J. Biol. Chem.* **243**:6234–6241.

Bock, E., and Dissing, J., 1975, Demonstration of enolase activity connected to the brain-specific protein 14-3-2. *Scand. J. Immunol.* **4**:31–36.

Bock, E., Fletcher, L., Rider, C. C., and Taylor, C. B., 1978, The nature of the two proteins of brain specific antigen 14-3-2, *J. Neurochem.* **30**:181–185.

Caravatti, M., Perriard, J. C., and Eppenberger, H. M., 1979, Developmental regulation of creatine kinase isoenzymes in myogenic cell cultures from chicken, *J. Biol. Chem.* **254**:1388–1394.

Chan, L., and O'Malley, B. W., 1976, Mechanism of action of the sex steroid hormones, *N. Engl. J. Med.* **294:**1322–1328, 1372–1381, and 1430–1437.

Cicero, T. J., Cowan, W. N., and Moore, B. W., 1970a, Changes in the concentrations of the two brain specific proteins, S-100 and 14-3-2 during the development of the avian optic tectum, *Brain Res.* **24:**1–10.

Cicero, T. J., Cowan, W. M., Moore, B. W., and Suntzeff, V., 1970b, Cellular localization of the two brain specific proteins S-100 and 14-3-2, *Brain Res.* **18:**25–34.

Cidlowski, J. A., and Muldoon, T. C., 1978, The dynamics of intracellular estrogen receptor regulation as influenced by 17β-estradiol, *Biol. Reprod.* **18:**234–246.

Cleveland, D. A., Fischer, S. G., Kirschner, M. W., and Laemmli, U. K., 1977, Peptide mapping by limited proteolysis in sodium dodecyl sulfate and analysis by gel electrophoresis, *J. Biol. Chem.* **252:**1102–1106.

Cohen, S., O'Malley, B. W., and Stastny, M., 1970, Estrogenic induction of ornithine decarboxylase *in vivo* and *in vitro, Science* **170:**336–338.

De Angelo, A. B., and Gorski, J., 1970, Role of RNA synthesis in the estrogen induction of a specific uterine protein, *Proc. Natl. Acad. Sci. USA* **66:**693–700.

de Asua, J., Rozengurt, L., and Carminatti, H., 1968, Estradiol induction of pyruvate kinase in the rat uterus, *Biochim. Biophys. Acta* **170:**254–262.

Donahue, T. M. Jr., Mahowald, T. A., and Barker, K. L., 1977, Identification and labeling of the NH₂-terminal residue of uterine glucose-6-phosphate dehydrogenase, *Fed. Proc. Fed. Am. Soc. Exp. Biol.* **36:**799.

Donahue, T. M. Jr., Mahowald, T. A., and Barker, K. L., 1979, Post-transcriptional control of glucose-6-phosphate dehydrogenase synthesis in the uterus by estradiol. *Fed. Proc. Fed. Am. Soc. Exp. Biol.* **38:**329.

Dupont-Mairesse, N., and Galand, P., 1975, Estrogen action: Induction of the synthesis of a specific protein (IP) in the myometrium, the stroma and the luminal epithelium of the rat uterus, *Endocrinology* **96:**1587–1591.

Eckstein, B., and Villee, C. A., 1966, Effects of estradiol on enzymes of carbohydrate metabolism in rat uterus, *Endocrinology* **78:**409–411.

Eppenberger, H. M., Dawson, D. M., and Kaplan, N. O., 1967, The comparative enzymology of creatine kinases. I. Isolation and characterization from chicken and rabbit tissues, *J. Biol. Chem.* **242:**204–209.

Focant, B., and Watts, D. C., 1973, Properties and mechanism of action of creatine kinase from ox smooth muscle, *Biochem. J.* **135:**265–276.

Galand, P., Flandroy, L., and Mairesse, N., 1978, Relationship between the estrogen induced protein IP and other parameters of estrogenic stimulation. A hypothesis, *Life Sci.* **22:**217–238.

Gorski, J., Sarff, M., and Clark, J., 1971, The regulation of uterine concentration of estrogen binding protein, *Adv. Biosci.* **7:**5–20.

Gulino, A., Sumida, C., Gelly, C., Giambiagi, N., and Pasqualini, J. R., 1981, Comparative dynamic studies on the biological responses to estriol and 17β estradiol in the fetal uterus of guinea pig: Relationship to circulating estrogen concentrations, *Endocrinology* **109:**748–756.

Harris, J. N., and Gorski, J., 1978a, Estrogen stimulation of DNA dependent DNA polymerase activity in immature rat uterus, *Mol. Cell. Endocrinol.* **10:**293–305.

Harris, J., and Gorski, J., 1978b, Evidence for a discontinuous requirement for estrogen in stimulation of deoxyribonucleic acid synthesis in the immature rat uterus, *Endocrinology* **103:**240–245.

Higgins, S. J., and Gehring, U., 1978, Molecular mechanisms of steroid hormone action, *Adv. Cancer Res.* **28:**313–397.

Iacobelli, S., Longo, P., and Ranelletti, F., 1979, Measurement of a specific estrogen induced protein (IP) in rat and mouse tissues, in: *Research on Steroids.* Vol. 8 (A. Klopper, L. Lerner, S. J. Van der Molen, and F. Sciarra, eds.), pp. 53–56, Academic Press, New York.

Iacobelli, S., Marchetti, P., Bartoccioni, E., Natoli, V., Scambia, G., and Kaye, A. M., 1981, Steroid-induced proteins in human endometrium, *Mol. Cell. Endocrinol.* **23:**321–331.

Iyengar, M. R., and Iyengar, C. W. L., 1980, Characterization of uterine muscle creatine kinase response to estrogen, *Fed. Proc. Fed. Am. Soc. Exp. Biol.* **39:**2171.

Iyengar, M. R., Fluellen, C. E., and Iyengar, C. W. L., 1980, Increased creatine kinase in the hormone-stimulated smooth muscle of the bovine uterus, *Biochem. Biophys. Res. Commun.* **94:**948–954.

Jacobus, W. E., and Lehninger, A. L., 1973, Creatine kinase of rat heart mitochondria. Coupling of creatine phosphorylation to electron transport, *J. Biol. Chem.* **248:**4803–4810.

Jellinck, P. H., Newcombe, A., and Keeping, H. S., 1979, Peroxidase as a marker enzyme in estrogen-responsive tissues, *Adv. Enzyme Regul.* **17:**325–342.

Katz, J., Finlay, T. H., Tom, C., and Levitz, M., 1980, A new hormone-responsive hydrolase activity in the mouse uterus, *Endocrinology* **107:**1725–1730.

Katzenellenbogen, B. S., and Gorski, J., 1972, Estrogen action *in vitro:* Induction of the synthesis of a specific uterine protein, *J. Biol. Chem.* **247:**1299–1305.

Katzenellenbogen, B. S., and Gorski, J., 1975, Estrogen actions on syntheses of macromolecules in target cells, in: *Biochemical Actions of Hormones.* Vol. 3 (G. Litwak, ed.), pp. 187–243, Academic Press, New York.

Katzenellenbogen, B. S., and Greger, N. G., 1974, Ontogeny of uterine responsiveness to estrogen during early development in the rat, *Mol. Cell. Endocrinol.* **2:**31–42.

Katzenellenbogen, B. S., and Leake, R. E., 1974, Distribution of tbe oestrogen-induced protein and of total protein between endometrial and myometrial fractions of the immature and mature rat uterus, *J. Endocrinol.* **63:** 439–449.

Katzenellenbogen, B. S., and Tsai, T-L. S., Tatee, T., and Katzenellenbogen, J. A., 1979, Estrogen and antiestrogen action. Studies in reproductive target tissues and tumors, *Adv. Exp. Med. Biol.* **117:**111–132.

Katzman, P. A., Larson, D. L., and Podratz, K. C., 1971, Effects of estradiol on metabolism of vaginal tissue, in: *The Sex Steroids: Molecular Mechanisms* (K. W. McKerns, ed.), pp. 107–147, Appleton-Century-Crofts, New York.

Kaye, A. M., and Reiss, N., 1980, The uterine "estrogen induced" protein (IP): Purification, distribution and possible function, in: *Steroid Induced Uterine Proteins* (M. Beato, ed.), pp. 3–19, Elsevier/North Holland, Amsterdam.

Kaye, A. M., Icekson, I., and Lindner, H. R., 1971, Stimulation by estrogens of ornithine and S-adenosylmethionine decarboxylases in the immature rat uterus, *Biochim. Biophys. Acta* **252:**150–159.

Kaye, A. M., Sheratzky, D., and Lindner, H. R., 1972, Kinetics of DNA synthesis in immature rat uterus: Age dependence and estradiol stimulation, *Biochim. Biophys. Acta* **261:**475–486.

Kaye A. M., Icekson, I., Lamprecht, S. A., Gruss, R., Tsafriri, A., and Lindner, H. R., 1973, Stimulation of ornithine decarboxylase activity by luteinizing hormone in immature and adult rat ovaries, *Biochemistry* **12:**3072–3076.

Kaye, A. M., Sömjen, D., King, R. J. B., Sömjen, G., Icekson, I., and Lindner, H. R., 1974, Sequential gene expression in response to estradiol-17β during post-natal development of rat uterus, *Adv. Exp. Med. Biol.* **44:**383–402.

Kaye, A. M., Sömjen, D., Sömjen, G., Walker, M., Icekson, I., and Lindner, H. R., 1975a, Regulation of macromolecular synthesis by oestrogen: a developmental approach, *Biochem. Soc. Trans.* **3:**1151–1156.

Kaye, A. M., Walker, M. D., and Sömjen, D., 1975b, Separation of "estrogen-induced" protein from phosphoprotein phosphatase activity of immature rat uterus, *Proc. Natl. Acad. Sci. USA* **72:**2631–2634.

Kaye, A. M., Reiss, N., and Walker, M. D., 1980a, Sequential acquisition of responsiveness to estrogen in the rat uterus, in: *Development of Responsiveness to Steroid Hormones* (A. M. Kaye and M. Kaye, eds.), Oxford, Pergamon Press.

Kaye, A. M., Reiss, N., Iacobelli, S., Bartoccioni, E., and Marchetti, P., 1980b, The "estrogen-induced protein" in normal and neoplastic cells, in: *Progress in Cancer Research and Therapy*

Vol. 14 (S. Iacobelli, R. J. B. King, H. R. Lindner, and M. E. Lipman, eds.), pp. 41–52, Raven Press, New York.

Kaye, A. M., Reiss, N., Shaer, A., Sluyser, M., Iacobelli, S., Amroch, D., and Soffer, Y., 1981, Estrogen responsive creatine kinase in normal and neoplastic cells, *J. Steroid Biochem.* **15**:69–75.

Kessler, S. W., 1975, Rapid isolation of antigens from cells with a staphylococcal protein A-antibody adsorbent: Parameters of the interaction of antibody-antigen complexes with protein A, *J. Immunol.*, **115**:1617–1624.

King, R. J. B., Sömjen, D., Kaye, A. M., and Lindner, H. R., 1974, Stimulation by oestradiol-17β of specific cytoplasmic and chromosomal protein synthesis in immature rat uterus. *Mol. Cell. Endocrinol.* **1**:21–36.

Kirby, K. S., 1968, Isolation of nucleic acids with phenolic solvents, *Methods Enzymol.* **12B**:87–97.

Kirkland, J. L., Barrett, G. N., and Stancel, G. M., 1981a, Decreased cell division of the luminal epithelium of diabetic rats in response to 17β-estradiol, *Endocrinology* **109**:316–318.

Kirkland, J. L., Gardner, R. M., Mukku, V. R., Akhtar, M., and Stancel, G. M., 1981b, Hormonal control of uterine growth: The effect of hypothyroidism on estrogen-stimulated cell division, *Endocrinology* **108**:2346–2351.

Korach, K. S., Harris, S. E., and Carter, D. B., 1981, Uterine proteins influenced by estrogen exposure: Analysis by two-dimensional gel electrophoresis. *Mol. Cell. Endocrinol.* **21**: 243–254.

Laemmli, U. K., 1970, Cleavage of structural proteins during the assembly of the head of bacteriophage T4, *Nature* **227**:680–685.

Lawson, G. M., Knoll, B. J., Anderson, J. N., Vanderbilt, J. N., Tsai, M. J., Woo, S. L. C., and O'Malley, B. W., 1981, Studies on the chromosomal structure of hormone inducible genes, Proceedings of the 63rd Annual Meeting, The Endocrine Society, *Endocrinology* **108**(suppl.):172 (abst.).

Leavitt, W. W., Chen, T. J., and Allen, T. C., 1977, Regulation of progesterone receptor formation by estrogen action, *Ann. N.Y. Acad. Sci.*, **286**:210–225.

Levy, H. R., 1979, Glucose-6-phosphate dehydrogenase, *Adv. Enzymol.* **48**:97–192.

Luh, W., and Henkel, E., 1965, Phosphotransferasen in menschlicher skelett-und utersmuskulatur, *Z. Geburtshilfe Gynakol.*, **163**:279–288.

Lyttle, C. R., and De Sombre, E. R., 1977, Uterine peroxidase as a marker for estrogen action, *Proc. Natl. Acad. Sci. USA* **74**:3162–3166.

Lyttle, C. R., Garay, R. V., and De Sombre, E. R., 1979, Ontogeny of the estrogen inducibility of uterine peroxidase, *J. Steroid Biochem.* **10**:359–363.

Mairesse, N., and Galand, P., 1979, Comparison between the action of estradiol and that of the antiestrogen U-11,100 A on the induction in the rat uterus of a specific protein (the induced protein), *Endocrinology* **105**:1248–1253.

Mairesse, N., and Galand, P., 1982, Estrogen-induced proteins in luminal epithelium, endometrial stroma and myometrium of the rat uterus. *Mol. Cell. Endocrinol.*, **28**:671–679.

Mairesse, N., Heuson, J. C., Galand, P., and Leclercq, G., 1977, Oestrogen-induced protein in rat mammary tumor and uterus, *J. Endocrinol.* **75**:331–332.

Mairesse, N., Reiss, N., Galand, P., and Kaye, A. M., 1981, Nafoxidine responsive uterine protein: Further characterization and distinction from the "estrogen-induced" protein (IP), *Mol. Cell. Endocrinol.* **24**:53–63.

Malnick, S. D. H., and Kaye, A. M., 1982, Estrogen-responsive creatine kinase in immature rat ovary, *J. Steroid Biochem.* **17**:V.

Marangos, P. J., Zomzely-Neurath, C., and York, C., 1975a, Immunological studies of a nerve specific protein, *Arch. Biochem. Biophys.* **170**:289–293.

Marangos, P. J., Zomzely-Neurath, C., Luk, D. C. M., and York, C., 1975b, Isolation and characterization of the nervous system specific protein 14-3-2 from rat brain, *J. Biol. Chem.* **250**:1884–1891.

Marangos, P. J., Zomzely-Neurath, C., and York, C., 1976, Determination and characterization of neuron specific protein (NSP) associated enolase activity, *Biochem. Biophys. Res. Commun.* **68:**1309–1316.

Marangos, P. J., Zis, A. P., Clark, R. L., and Goodwin, F. K., 1978, Neuronal, non-neuronal and hybrid forms of enolase in brain: Structural, immunological and functional comparisons, *Brain Res.* **150:**117–133.

Mayol, R. F., and Thayer, S. A., 1970, Synthesis of estrogen-specific proteins in the uterus of the immature rat, *Biochemistry* **9:**2484–2489.

Miura, S., Burzio, L., and Koide, S. S., 1972, Studies on deoxyribonucleic acid synthesis in uteri of immature mouse, *Horm. Metab. Res.* **4:**273–277.

Moore, B. W., and Perez, V. J., 1968, Specific acidic proteins of the nervous system, in: *Physiological and Biochemical Aspects of Nervous Integration* (F. D. Carlson, ed.), pp. 343–360, Prentice-Hall, Engelwood Cliffs, New Jersey.

Mueller, G. C., Herranen, A. M., and Jervell, K. F., 1958, Studies on the mechanism of action of estrogens, *Rec. Prog. Horm. Res.* **14:**95–139.

Notides, A., and Gorski, J., 1966, Estrogen-induced synthesis of a specific uterine protein, *Proc. Natl. Acad. Sci. USA* **56:**230–235.

O'Farrell, P. H., 1975, High resolution two-dimensional electrophoresis of proteins, *J. Biol. Chem.* **250:**4007–4021.

Ozasa, H., Tominaga, T., Nishimura, T., and Takeda, T., 1981, Lysyl oxidase activity in the mouse uterine cervix is physiologically regulated by estrogen, *Endocrinology* **109:**618–621.

Peleg, S., De Boever, J., and Kaye, A. M., 1979, Replenishment and nuclear retention of oestradiol-17β receptors in rat uteri during postnatal development, *Biochim. Biophys. Acta* **587:**67–74.

Pelham, H. R. B., and Jackson, R. J., 1976, An efficient mRNA-dependent translation system from reticulocyte lysates, *Eur. J. Biochem.* **67:**247–256.

Pennequin, P., Robel, P., and Baulieu, E. E., 1975, Steroid-induced early protein synthesis in rat uterus and prostate, *Eur. J. Biochem.* **60:**137–145.

Raynaud, J. P., Moguilewsky, M., and Vannier, B., 1980, Influence of rat estradiol binding plasma protein (EBP) on estrogen binding to its receptor and on induced biological responses, in: *The Development of Responsiveness to Steroid Hormones* (A. M. Kaye and M. Kaye, eds.), Pergamon Press, Oxford. *Adv. in Biosci.* **25,** 59–75.

Reiss, N. A., and Kaye, A. M., 1981, Identification of the major component of the estrogen-induced protein of rat uterus as the BB isozyme of creatine kinase, *J. Biol. Chem.* **256:**5741–5749.

Schmidt, H., Noak, I., Walther, H., and Voigt, K. D., 1967, Einfluss von cyclus und exogener hormonzufuhr auf stoffwechsel vorgange des rattenuterus. *Acta Endocrinol.* **56:**231–243.

Schwark, W. S., Singhal, R. L., and Ling, G. M., 1969, Metabolic control mechanisms in mammalian systems. VIII. Estrogenic induction of fructose 1,6-diphosphate aldolase in the rat uterus. *Biochim. Biophys. Acta* **192:**106–117.

Scott, D. B. M., and Lisi, A. G., 1960, Changes in enzymes of the uterus of the ovariectomized rat after treatment with oestradiol, *Biochem. J.* **77:**52–63.

Segal, S. J., Scher, W., and Koide, S. S., 1977, Estrogens, nucleic acids and protein synthesis in uterine metabolism, in: *Biology of the Uterus,* 2nd Ed. (R. M. Wynn, ed.), pp. 139–201, Plenum Press, New York.

Shaer, A., Amroch, D., Malnick, S., and Kaye, A. M., 1982, Estrogen-responsive creatine kinase in normal human breast tissue, *Israel J. Med. Sci.* **18:**565.

Skipper, J. K., Eakle, S. D., and Hamilton, T. H., 1980, Modulation by estrogen of synthesis of specific uterine proteins, *Cell* **22:**69–78.

Smith, E. R., and Barker, K. L., 1974, Effects of estradiol and NADP on the rate of synthesis of uterine glucose-6-phosphate dehydrogenase, *J. Biol. Chem.* **249:**6541–6547.

Smith, E. R., and Barker, K. L., 1977, Effects of estradiol and NADP on the rate of degradation of uterine glucose-6-phosphate dehydrogenase, *J. Biol. Chem.* **252:**3709–3714.

Smith, D. E., and Gorski, J., 1968, Estrogen control of uterine glucose metabolism. An analysis based on the transport and phosphorylation of 2-deoxyglucose, *J. Biol. Chem.* **243:**4169–4174.

Sömjen, D., Sömjen, G., King, R. J. B., Kaye, A. M., and Lindner, H. R., 1973a, Nuclear binding of oestradiol-17β and induction of protein synthesis in the rat uterus during postnatal development, *Biochem. J.* **136:**25–33.

Sömjen, D., Kaye, A. M., and Lindner, H. R., 1973b, Postnatal development of uterine response to estradiol-17β in the rat, *Dev. Biol.* **31:**409–412.

Sömjen, G. J., Kaye, A. M., and Lindner, H. R., 1974, Oestradiol-17β binding proteins in the rat uterus: Changes during postnatal development, *Mol. Cell. Endocrinol.* **1:**341–353.

Sömjen, G. J., Kaye, A. M., and Lindner, H. R., 1976, Demonstration of 8S-cytoplasmic estrogen-receptor in rat Mullerian duct, *Biochim. Biophys. Acta* **428:**787–791.

Sonnenschein, C., and Soto, A. M., 1978, Pituitary uterotrophic effect in the estrogen-dependent growth of the rat uterus, *J. Steroid Biochem,* **9:**533–537.

Stormshak, F., Harris, J. N., and Gorski, J., 1978, Nuclear estrogen receptor and DNA synthesis, in: *Receptors and Hormone Action,* Vol. II (B. W. O'Malley and L. Birnbaumer, eds.), pp. 63–81, Academic Press, New York.

Strohman, R. C., Moss, P. S., Micou-Eastwood, J., Spector, D., Przbyla, A., and Patterson, B., 1977, Messenger RNA for myosin polypeptides: Isolation from single myogenic cell cultures, *Cell* **10:** 265–273.

Tam, A., and Spaziani, E., 1970, The effect of estradiol 17β on uterine adenosine triphosphatases of the rat, *Fed. Proc. Fed. Am. Soc. Exp. Biol.* **29:**249.

Teng, C-S, and Hamilton, T. H., 1970, Regulation by estrogen of organ-specific synthesis of a nuclear acidic protein, *Biochem. Biophys. Res. Commun.* **40:**1231–1238.

Thomas, P. J., and Knight, A., 1978, Sexual differentiation of the brain, in: *Current Studies of Hypothalamic Function,* Vol. I (K. Lederis and W. L. Veale, eds.), pp. 192–203, Karger, Basel.

Tremblay, G. C., and Thayer, S. A., 1964, The effect of estradiol-17β on the activity of carbamoyl phosphate: L-aspartate carbamoyl transferase in the uteri of immature rats, *J. Biol. Chem.* **239:**3321–3324.

Valadares, J. R. E., Singhal, R. L., and Parulekar, M. R., 1968, 17β-Estradiol: Inductor of uterine hexokinase, *Science* **159:**990–991.

Vertes, M., Kornyei, J., Nagy, L., Vertes, Z., and Kovacs, S., 1981, Stimulation by oestradiol of soluble-protein synthesis in rat hypothalamus. *Mol. Cell. Endocrinol.* **22:**329–338.

Vokaer, A., Iacobelli, S., and Kram, R., 1974, Phosphoprotein phosphatase activity associated with estrogen-induced protein in rat uterus. *Proc. Natl. Acad. Sci USA* **71:**4482–4486.

Walker, M. D., and Kaye, A. M., 1981, mRNA for the rat uterine estrogen induced protein: Translation *in vitro* and regulation by estrogen, *J. Biol. Chem.* **256:**23–26.

Walker, M. D., Gozes, I., Kaye, A. M., Reiss, N., and Littauer, U. Z., 1976, The "estrogen induced protein": Quantitation by autoradiography of polyacrylamide gels, *J. Steroid Biochem.* **7:**1083–1085.

Walker, M. D., Kaye, A. M., and Fridlender, B. R., 1978, Age-dependent stimulation by estradiol 17β of DNA polymerase α in immature rat uterus, *FEBS Lett.* **92:**25–28.

Walker, M. D., Negreanu, V., Gozes, I., and Kaye, A. M., 1979, Identification of the "estrogen-induced protein" in uterus and brain of untreated immature rats, *FEBS Lett.* **98:**187–191.

Weissbach, A., 1977, Eukaryotic DNA polymerases, *Annu. Rev. Biochem.* **46:**25–47.

Wood, T., 1963, Adenosine triphosphate-creatine phosphotransferase from ox brain: Purification and isolation, *Biochem. J.* **87:**453–462.

Wood, T., 1964, The purification of enolase from cerebral tissue, *Biochem. J.* **91:**453–460.

7

Estrogen-Induced Uterine Hypertrophy

John T. Knowler

1. Introduction

The mode of action of estrogen has been most studied in those tissues, such as the oviduct and the avian or amphibian liver, where its presence induces the synthesis of major secreted proteins. In the uterus, however, the effects of estrogens are not marked by the preferential synthesis of large quantities of a small number of proteins. Rather, the hormone induces an overall growth; a sequence of hypertrophy followed by hyperplasia which initiates the process whereby the tissue is prepared for a possible pregnancy. The entire growth process is characterized by a sequential stimulation of the synthesis of all classes of RNA, total protein, and DNA but the hypertrophic phase of the response appears to depend in large measure on the mobilization of existing ribosomes and the estrogen-induced synthesis of new ribosomes. This review summarizes our current knowledge of the events leading up to, and apparently prerequisite for, this activation process.

2. Estradiol-17β-Stimulated Uterine Hypertrophy

Uterine hypertrophy begins with the stimulated synthesis of uterine RNA. A number of reports in the past have suggested that this resulted wholly or in

John T. Knowler ● Department of Biochemistry, University of Glasgow, Glasgow G12 8QQ, Scotland.

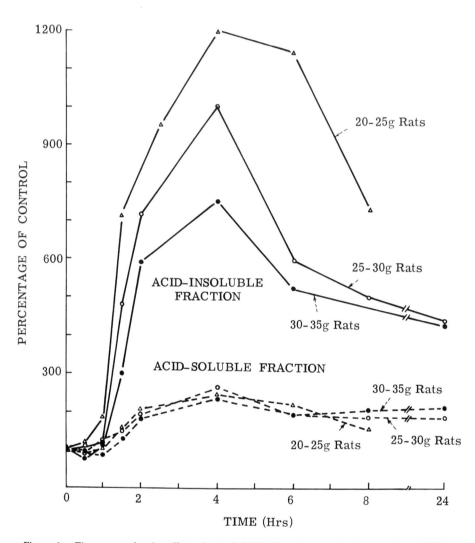

Figure 1. Time-course for the effect of estradiol-17β (1 μg/rat) on the incorporation of RNA precursor into 18- to 21-day-old rat uteri. Hormone was administered intraperitoneally and 10 μCi each of [5-³H]uridine and [8-³H]guanosine was given intravenously 30 min before death. Results were calculated as dpm/μg of DNA and expressed as percentages of the uptake in non-hormone-treated controls. Each point represents a mean of at least three separate experiments, △, 20- to 25-g rats; ◯, 25- to 30-g rats; ●, 30- to 35-g rats. (Reproduced from Knowler *et al.* (1975) by courtesy of the editorial board of the Biochemical Society.)

part from increases in the pool sizes of precursors (Miller and Baggett, 1972). There is no doubt that such increases in pool sizes do occur (Oliver and Kellie, 1970) but probably only as a result of estrogen-induced imbibition of water (Oliver, 1971). Furthermore, our studies of the entry of RNA precursor into the acid-soluble and acid-insoluble components of the uterus clearly show that the stimulated synthesis of uterine RNA cannot be accounted for by increased pools of precursor (Fig. 1). Stimulation of RNA synthesis peaks at 6–12 times that of controls, 2–4 hr after the administration of 0.3–1.0 μg of estradiol-17β per 3-week-old rat. The degree of stimulation varies with the weight of the rats (Fig. 1) and is also dependent on the route of precursor administration (Knowler and Smellie, 1971; Knowler *et al.,* 1975). Conversely, pool sizes of precursor never increase to more than 2.5 times control levels (Fig. 1).

The RNA-synthesis phase of uterine hypertrophy is followed by stimulation of total protein synthesis but this is a simplification of a more complex series of interdependent events illustrated in Fig. 2. In the immature rat uterus, responding to a single injection of estradiol-17β, stimulated hnRNA synthesis is first seen 30 min after administration. The newly made hnRNA matures to mRNA and, from 1 hr after hormone injection, causes the aggregation of preexisting ribosomes into polysomes. Stimulated rRNA synthesis can be detected at $1\frac{1}{2}$ hr and peaks at 2–4 hr. However, incorporation of the newly made rRNA into ribosomes does not peak until 12 hr after hormone treatment. Each event in the scheme is triggered in due sequence, reaches a maximum, and then falls off. If more metabolically

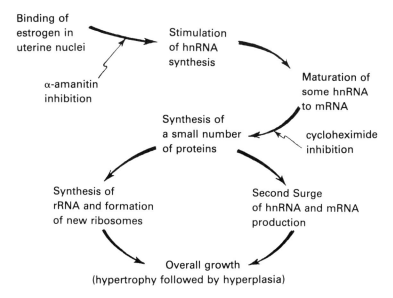

Figure 2.   Sequence of events in estrogen-induced growth of the uterus of the immature rat.

stable estrogen derivatives or estrogen in parafin wax implants are employed, responses are sustained to a much greater extent and if ovariectomized animals are used the time course is condensed in a manner analagous to secondary stimulation of the avian oviduct.

3. Estrogen-Stimulated hnRNA Synthesis and Its Maturation to mRNA

The earliest transcriptional response to estrogen is the stimulated synthesis of hnRNA. This was first detected *in vivo* as the increased incorporation of precursor into high-mol.-wt. RNA species which were identified as hnRNA on the basis of base composition, a low degree of methylation, a nuclear location, and rapid synthesis and decay (Knowler and Smellie, 1973). Stimulated hnRNA

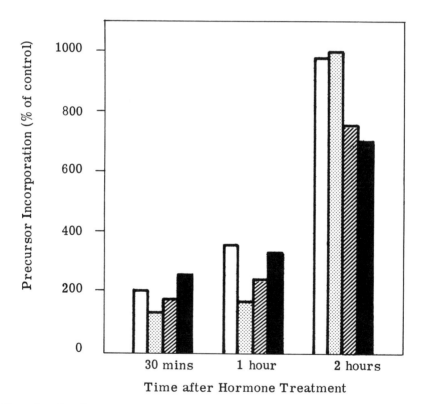

Figure 3.   The effect of estradiol-17β on the synthesis of uterine hnRNA. High-mol.-wt. uterine hnRNA was prepared and fractionated on poly "U" sepharose as described by Aziz and Knowler, 1978. □, total high-mol.-wt. hnRNA; ▨, nonadenylated high-mol.-wt. hnRNA; ▧, oligoadenylated, high-mol.-wt. hnRNA; ■, polyadenylated, high-mol.-wt. hnRNA.

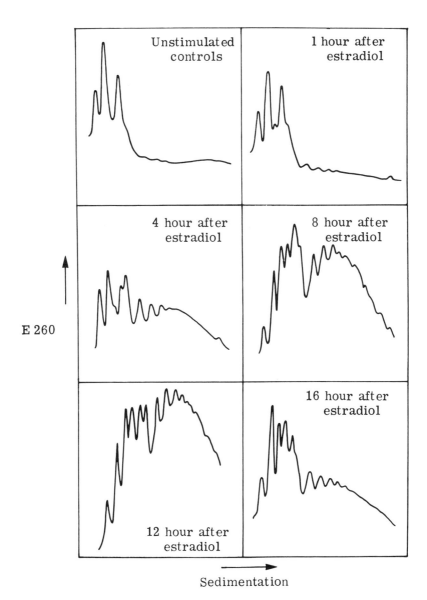

Figure 4. Polyribosomal profiles of the immature rat uterus responding to estradiol-17β. Polyribosomes were prepared as described by Merryweather and Knowler (1980). (Adapted from Merryweather and Knowler (1980) by courtesy of the editorial board of the *Biochemical Journal*.)

synthesis was also demonstrated *in vitro* by the increased activity of RNA poly-merase II in nuclei isolated from estrogen-treated ovariectomized rats and im-mature rabbits (Glasser *et al.*, 1972; Borthwick and Smellie, 1975; Webster and Hamilton, 1976). These early results indicated that the stimulated synthesis of hnRNA could be detected as early as 30 min after the administration of estrogen and this was confirmed when high-mol.-wt. uterine hnRNA was exhaustively purified through nondenaturing and denaturing sucrose density gradients. Figure 3 shows that, by 30 min after estradiol-17β administration, hnRNA synthesis was twice that of untreated controls and, by 2 hr posttreatment it was 10-fold (Aziz and Knowler, 1978). The highly purified high-mol.-wt. hnRNA could be fractionated into polyadenylated, oligoadenylated, and nonpolyadenylated spe-cies by differential elution from poly "U" Sepharose. Species not retained by the affinity gel were found to be nonadenylated, those eluted by 15% formamide were oligoadenylated, and those eluted by 90% formamide were polyadenylated. Examination of the effect of estradiol on the synthesis of each type of molecule revealed that the synthesis of all classes of hnRNA was stimulated. However, the synthesis of polyadenylated hnRNA accounted for most of the early response

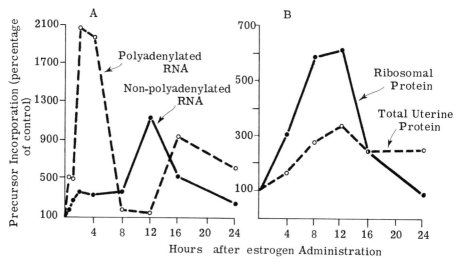

Figure 5.   Effects of estradiol-17β on the incorporation of precursor into the components of uterine polysomes. Groups of 12 female rats aged 18–21 days each received 1 μg of estradiol-17β or carrier at various times before death. RNA synthesis was monitored by the incorporation of [5,6-³H]uridine administered intravenously at 100 μCi/rat 30 min before death, and protein synthesis was monitored as the incorporation of [4,5-³H]leucine which was administered intravenously 1 hr before death. Polyadenylated and nonpolyadenylated RNA, total protein, and ribosomal protein were purified as described by Merryweather and Knowler, 1980. (Reproduced from Merryweather and Knowler (1980) by courtesy of the editorial board of the *Biochemical Journal*.)

while synthesis of nonadenylated species was most dramatic at later times (Fig. 3) (Aziz and Knowler, 1978).

The maturation of cytoplasmic mRNA from hnRNA could be followed kinetically by its appearance in extractable hnRNP particles (Knowler, 1976), by the detection of mRNA sequences in the hnRNA (Aziz and Knowler, 1980), and then by the entry into the cytoplasm of newly synthesized mRNA (Merryweather and Knowler, 1980). The ribosomes of the immature rat uterus are largely in the form of monosomes (Fig. 4). By 1 hr after a single estradiol-17β administration, polysomes are beginning to form and these continue to accumulate and increase in average size up to 12 hr after hormone administration, after which they decrease in size. Incorporation of radioactive precursor reveals (Fig. 5) that the early formation of polysomes results from the accumulation of preexisting ribosomes on newly synthesized mRNA. Only at later stages in the hormonal response are newly synthesized ribosomes incorporated into the polysomes (Fig. 5) (Merryweather and Knowler, 1980).

4. Studies on the mRNA Population of the Rat Uterus

$R_o t$ hybridization is a powerful technique with which to determine both the number of mRNA species in a population and changes within that population. We have derived estimates of the number of mRNAs in rat uterus by two $R_o t$ hybridization methods. Firstly, we have used reverse transcriptase to make cDNA copies of total polyadenylated polysomal mRNA and have used the kinetics of hybridization of the cDNA to its template, compared with the similar hybridization of a standard, to derive estimates of mRNA population complexity as described by Bishop *et al.* (1974), Young *et al.* (1974), and others. Secondly, we have conducted hybridizations between polysomal mRNA and unique genomic DNA (Galau *et al.*, 1974). To achieve this, polysomal polyadenylated mRNA was mercurated and hybridized to unique DNA which had been labeled to a high specific activity by nick translation. Hybrids were detected and quantified by their affinity for thiol-Sepharose.

An immature rat uterus, responding to a single injection of estradiol-17β, exhibits a peak of RNA synthesis 4 hr after administration. At this time, analysis of the mRNA population by cDNA hybridization reveals 8000 different polyadenylated mRNA species divided into three different abundance classes (Table I). Conversely, analysis by unique DNA–mRNA hybridization yields values of 12,000 sequences (Aziz *et al.*, 1979a). The fact that the latter method gives higher estimates of mRNA complexity than the former is well documented (see Aziz *et al.*, 1979a) and may be caused by cDNA hybridizations yielding underestimates of the numbers of the least abundant sequences. Regardless of which of these figures was correct, however, analysis of changes in mRNA complexity made it clear that estrogen administration caused considerable changes

Table I. Polyadenylated mRNA in the Uterus of Immature Rats Responding to
Estradiol-17β[a]

	Number of diverse sequences		
Source of uterine mRNA	Abundant	Intermediate abundant	Rare
Total polyadenylated mRNA species present 4 hr after estradiol administration	9	150	7800
Polyadenylated mRNA species present at 4 hr but not at 2 hr after estradiol administration	1–2	150	2530

[a] The data represent a computer analysis of the kinetics of hybridization of uterine mRNA to complementary cDNA. It is derived from data previously presented by Aziz *et al.* (1979a,b).

in the abundance of many species. Such changes were detected by heterologous hybridization. A cDNA prepared against an mRNA population, for example, of estrogen-treated rat uterus, was hybridized to the population from a different source, for example, uterine mRNA from untreated rats. The extent and kinetics of hybridization are a measure of the similarity between the two mRNA populations.

Figure 6 shows the results obtained when a cDNA copy of the uterine mRNA population of rats responding to 4 hr estrogen treatment was hybridized to its own template of 4 hr polyadenylated mRNA or to the mRNA population from 2-hr treated animals or to mRNA isolated from untreated animals (Aziz *et al.*, 1979b). The low abundance of polysomes in untreated animals (Fig. 4) makes the preparation of their mRNA very difficult and not enough could be prepared for a complete R_0t curve. It can be seen, however, that what polyadenylated mRNA could be prepared would only hybridize to 11% of the 4-hr cDNA. Animals receiving estradiol 2 hr before death contained uterine mRNA sequences able to saturate 56% of the 4-hr cDNA (Fig. 6). This implied that a high proportion of the 4-hr and 2-hr polyadenylated mRNA population was common but the slow rate of hybridization, relative to the homologous reaction, revealed that the common sequences were present in lower abundance in the animals receiving 2-hr hormone treatment. Furthermore, 24% of the cDNA did not hybridize to the 2-hr mRNA by a R_0t of 10,000 mol·sec·liter^{-1}. This difference proved reproducible over numerous repeats involving several preparations and it is reasonable to assume that it represents mRNA species which are synthesized during the second 2 hr of the hormonal response but not during the first 2 hr. As such, these sequences are likely to be of considerable importance in the increased production of ribosomes initiated at this time (Aziz *et al.*, 1979b).

To investigate this change in mRNA population further, the cDNA derived from uterine polysomal polyadenylated mRNA of rats treated for 4 hr with estrogen was used to prepare fractions enriched in abundant and rare sequences, respectively. For the isolation of abundant cDNA sequences, total cDNA was

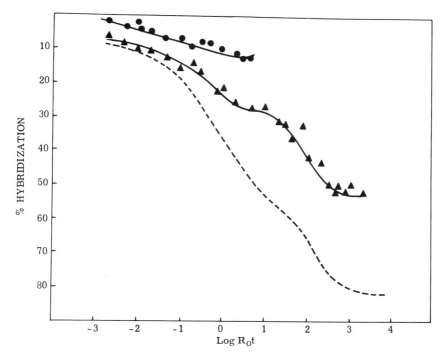

Figure 6.   Kinetics of hybridization of cDNA with homologous and heterologous mRNA. cDNA prepared from uterine mRNA of rats which had received 4-hr estrogen treatment was hybridized to: uterine mRNA from untreated immature rats (●———●), uterine mRNA from immature rats receiving 2-hr estrogen treatment (▲———▲, and mRNA from immature rats receiving 4-hr estrogen treatment (– – – –). (Reproduced from Aziz *et al.* (1979b) by courtesy of the editorial office of the *European Journal of Biochemistry.*)

hybridized to its own template to a R_0t of 1 mol·sec·liter^{-1} and double-stranded sequences were recovered by fractionation on hydroxyapatite. The cDNA was then purified from the hybrids by alkali digestion of the RNA, neutralization, and removal of salt. Rare sequence cDNAs were similarly isolated as cDNA which remained single stranded at a R_0t of 12 mol·sec·liter^{-1} (Aziz *et al.*, 1979b). The cDNA fractions were employed in further homologous and heterologous hybridization analyses of the differences between the uterine polyadenylated mRNA populations of rats responding to 2 and 4 hr of estradiol exposure.

There was considerable sequence homology between the most abundant polyadenylated mRNA sequences of 4-hr and 2-hr treated animals (Fig. 7A). Such findings were supported when uterine polysomal mRNA was translated in *in vitro* protein synthesizing systems. Detectable translation products in these systems reflect proteins synthesized from the most abundant mRNA species and those made from uterine mRNA isolated from 2- or 4-hr treated animals are very similar. Indeed, the translation products of uterine polysomal mRNA, detected

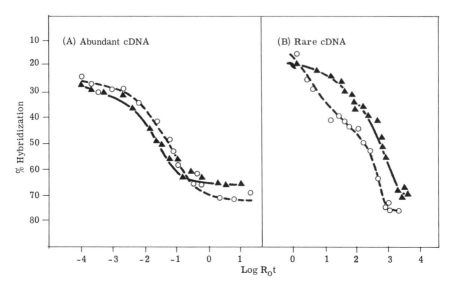

Figure 7. Kinetics of hybridization of fractionated cDNA with homologous and heterologous mRNA. cDNA prepared from uterine mRNA of rats which had received estradiol-17β 4 hr before death was fractionated as described by Aziz *et al.* (1979b) and the fractions were hybridized to: uterine mRNA from immature rats receiving 2-hr estrogen treatment (▲———▲), uterine mRNA from rats receiving 4-hr estrogen treatment (○– – –○). (Adapted from Aziz *et al.* (1979b) by courtesy of the editorial office of the *European Journal of Biochemistry*.)

as [^{35}S]methionine-labeled spots on two-dimensional protein fractionations (O'Farrell, 1975), are very similar at all stages of estrogen-induced development indicating that there are few qualitative changes in the abundant mRNA populations (Beaumont and Knowler, 1981). This situation not only contrasts with other estrogen target tissues, such as the fowl oviduct and the avian and amphibian liver, but also with androgen target tissues such as the prostate gland where techniques similar to those above led to the detection of four abundant proteins and their mRNA synthesized in response to the hormone (Parker and Mainwaring, 1977).

Figure 7B shows that there was also considerable, though incomplete, homology between the rare polyadenylated mRNA sequences of the 2-hr and 4-hr post-hormone-treated rat uterus. The figure also shows, however, that the preparation of rare cDNA was contaminated with sequences of intermediate abundance and that these hybridized incompletely to the 2-hr mRNA preparation. This led us to suspect that the difference between the 2-hr and 4-hr mRNA populations resided primarily in species of intermediate abundance. Despite many attempts, however, we were unable to prepare a cDNA fraction sufficiently enriched in intermediate abundant sequences to test this possibility by the above methodology. As an alternative approach, which did allow us to quantitate the difference, the following experiment was performed. A preparation of poly-adenylated mRNA from 2-hr treated rat uteri was hybridized to a R_0t of 2000 mol·sec·liter^{-1} to total cDNA prepared from mRNA of 4-hr treated rats. The cDNA sequences that remained single-stranded were assumed to represent the sequences present at 4 hr but not 2 hr after hormone administration. They were isolated by hydroxyapatite chromatography and rehybridized; this time to the 4-hr mRNA population (Aziz *et al.*, 1979b). Table I summarizes the numerical interpretation of the hybridization curve obtained and compares the numbers of

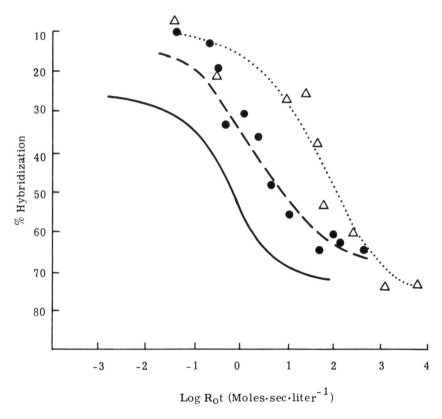

Figure 8.   Hybridization of cDNA of abundant uterine mRNA against uterine polyadenylated hnRNA. The solid curve (———) represents the hybridization of abundant cDNA to its own template: namely polysomal mRNA from rats receiving 1 μg estradiol-17β 4 hr before death. The remaining curves illustrate heterologous hybridizations between the cDNA and polyadenylated hnRNA: hnRNA recovered from 30-min estradiol-treated rats ($\triangle$ ••••• $\triangle$); hnRNA recovered from 2-hr estradiol-treated rats ($\bullet$– –$\bullet$). In each of these two curves, the curve itself is derived from hnRNA prepared under nondenaturing conditions while the data points represent a repeat experiment in which the hnRNA was prepared under denaturing conditions. (Reproduced from Aziz and Knowler (1980) by courtesy of the editorial board of the *Biochemical Journal*.)

mRNA species made between 2 and 4 hr after estrogen administration with the total population of 4-hr treated rats. It can be seen that virtually all of the intermediate abundant sequences present in the uteri of 4-hr treated animals were first made, or first appeared in that fraction, between 2 and 4 hr after hormone administration.

That the above changes in mRNA populations resulted from transcriptional events was indicated by the detection of similar changes in the abundance of sequences complimentary to mRNA within the nuclear hnRNA (Aziz and Knowler, 1980). Figure 8 shows the results obtained when cDNA, complementary to the abundant polysomal mRNAs of 4-hr estrogen-treated rats was hybridized to highly purified, polyadenylated, high-mol.-wt. hnRNA of 2-hr and 30-min treated rats. The data indicate that the abundant mRNA sequences make up 0.085% and 3.4%, respectively, of the 30-min and 2-hr polyadenylated hnRNA. The nonpolyadenylated hnRNA also contained mRNA sequences but at a much lower abundance. Thus in the 2-hr hnRNA, 0.073% of the preparation was complementary to the cDNA probe.

5. Estrogen-Stimulated rRNA and tRNA Synthesis

An understanding of the nature of the proteins encoded in the intermediate abundant mRNA species synthesized between 2 and 4 hr after estrogen administration to immature rats may be crucial to the investigation of uterine hypertrophy because it is at this time that the synthesis of new ribosomes is activated.

Stimulated rRNA synthesis in immature rats, receiving a single intraperitoneal injection of estradiol-17β, peaks at 2–4 hr after administration. This has been demonstrated by *in vivo* incorporation of precursor (Fig. 9) and by *in vitro* analysis of RNA polymerase I activity in nuclei isolated from the uteri of treated rats and rabbits (Glasser *et al.*, 1972; Borthwick and Smellie, 1975; Webster and Hamilton, 1976). Some evidence has been presented that the effect of the hormone might be in part explained by activated maturation of rRNA precursor molecules, possibly via increased activity of methylases (Knecht and Luck, 1977). However, it is clear that the main effect is at the level of transcription. Thus, the stimulated synthesis can be detected as a 10-fold increase in the incorporation of radioactive precursor into the 45 S preribosomal RNA and can be followed through the 32 S species into 28 S and 18 S rRNA (Knowler and Smellie, 1971).

That this estrogen-induced synthesis of rRNA depends on prior protein synthesis was first indicated when cycloheximide was found to inhibit the stimulation (Knowler and Smellie, 1971). Subsequent studies by Borthwick and Smellie (1975) employing a range of inhibitors, confirmed that the effect was dependent both on protein synthesis and on the initial hnRNA synthesis. The most telling

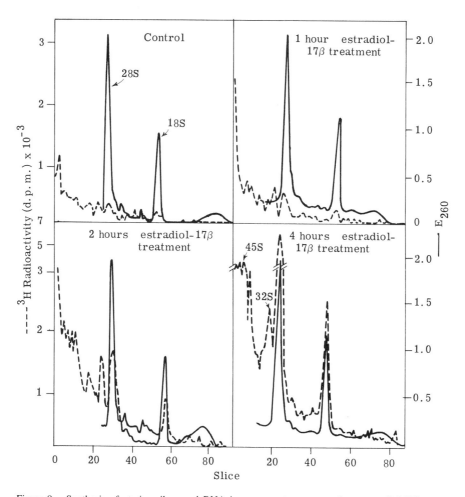

Figure 9. Synthesis of uterine ribosomal RNA in response to estrogen; 1 μg estradiol-17β was given by intraperitoneal injection to groups of eight 18–21 days-old rats weighing 25–30 g; 125 μCi/rat [5-³H]-uridine and [8-³H]-guanosine each was injected intravenously at various times before death. Purified RNA was fractionated on 2.7% polyacrylamide gels for 5 hr. For reasons of clarity the OD260 trace is omitted from the beginning of the gel. (Adapted from Knowler and Smellie (1971) by courtesy of the editorial board of the *Biochemical Journal.*)

of these studies employed α-amanitin, a specific inhibitor of RNA polymerase II. When this inhibitor was given 30 min prior to the administration of estrogen, not only was the activity of uterine polymerase II abolished but the stimulation of polymerase I was also abolished. The basal level of polymerase I was not, however, effected and, if α-amanitin was given 30 min after estradiol, at a time when hnRNA synthesis had already been stimulated to twice control levels, there was no inhibition of RNA polymerase I stimulation. Polymerase II was still totally inhibited (Borthwick and Smellie, 1975). In short, the stimulation of rRNA synthesis was dependent on the initial stimulation of hnRNA synthesis. Similar time-dependent responses to inhibitors of protein synthesis showed that this too was a prerequisite for stimulated rRNA synthesis and it would seem reasonable to conclude that the important proteins are transcribed from mRNA which matures from the hnRNA.

Perhaps slightly earlier than that of rRNA, estrogen also dramatically stimulates the synthesis of 5 S RNA and tRNA (Knowler and Smellie, 1971). Thus, by 2–4 hr after a single estradiol-17β injection, the synthesis of all stable RNA species has been activated and reached a maximum. Soon afterwards, there is a second surge of hnRNA synthesis (Borthwick and Smellie, 1975), this time presumably encoding the mRNA sequences of the structural proteins and enzymes to be synthesized during the hypertrophic phase of growth. Given all this, it is surprising that there follows a considerable delay between the nuclear events and their cytoplasmic continuation. Thus the accumulation of new ribosomes in the cytoplasm and the stimulation of protein synthesis does not peak until 12 hr after estradiol administration (Merryweather and Knowler, 1980).

6. The Nature of the Proteins on Which Stimulated Ribosome Production Might Depend

Our findings that stimulated uterine rRNA synthesis is dependent on hnRNA and protein synthesis are supported by similar observations both in the uterus (Raynaud-Jammet *et al.*, 1972; Nicollette and Babler, 1974) and in other systems (Schmid and Sekeris, 1973; Hadjiolov *et al.*, 1974; Lindell *et al.*, 1978) and it is tempting to speculate that the new mRNA, which appears during the first 4 hr after estrogen administration, and which comes to be dominated by sequences of intermediate abundance (Aziz *et al.*, 1979b), may have a crucial role in coding for proteins involved in the control of rRNA synthesis.

What then are these intermediate abundant mRNA sequences likely to code for? One possible protein fraction that might well be translated from them is that known as induced protein (IP) (Notides and Gorski, 1966) that has recently been shown to consist of creatine kinase (Reiss and Kaye, 1981) and an enolase isoenzyme (Kaye and Reiss, 1980). Replenishment of the cytoplasmic receptor protein is also initiated at this time (Korach and Ford, 1978) but none of these proteins are likely to be important in the control of ribosome synthesis.

6.1. RNA Polymerase

One obvious candidate protein is RNA polymerase I itself. It must be emphasized, however, that all of the studies of the uterine RNA polymerase described above measured enzyme activity in isolated nuclei. In such assays the effects of hormone on enzyme activity cannot be separated from effects on the chromatin template. Indeed, when either RNA polymerase I or II were extracted from isolated nuclei no effect of estrogen on their activity could be demonstrated (Borthwick and Smellie, 1975). It remains possible that the extraction process removed a hormone-modulated factor analogous to bacterial sigma factor but numerous attempts to demonstrate such a factor have been unsuccessful (Wells, 1973; Borthwick, 1974).

6.2. Ribosomal Proteins

Further possible candidate proteins that could be important in the controlled production of new ribosomes are the ribosomal proteins (r-proteins) themselves. We have followed the appearance of newly made r-proteins in uterine ribosomes by the incorporation of radioactive amino acids into proteins isolated from ribosomal subunits which had been rigorously purified free of nascent protein chains (Merryweather and Knowler, 1980). Since rRNA synthesis in response to a single administration of estradiol peaks after 2–4 hr while total protein synthesis does not peak until 12 hr, it might be expected that r-protein synthesis would precede that of total protein and mirror that of rRNA. Figure 5 shows that the opposite is the case. The appearance of newly made r-protein in ribosomes shows a very similar time course to that of total protein synthesis and so, surprisingly, does the incorporation of rRNA into new ribosomes. Thus, despite the fact that rRNA synthesis peaks at 10 times control levels 2–4 hr after estradiol-17β administration, incorporation into ribosomes is stimulated to only three times at this time. Conversely, at 12 hr after treatment, when rRNA synthesis is rapidly falling to control levels, incorporation into ribosomes reaches its maximum 11-fold stimulation. The implication is that between 2 and 12 hr after hormone treatment a large pool of unincorporated rRNA accumulates in uterine cells. There are indications that this pool of rRNA is substantially incorporated into ribosomes between 8 and 12 hr after estrogen administration as, at this time, the incorporation of radioactive protein is near maximal while that of newly made rRNA is only 3.5 times that of unstimulated controls. This may reflect the association of newly made protein with the pool of rRNA built up over the preceding 6 hr. It is hoped that further studies on cytoplasmic pools of rRNA and r-proteins and on the *in vitro* synthesis of ribosomal proteins on uterine polysomes may clarify these issues.

6.3. Ribonucleases and Ribonuclease Inhibitors

Another group of proteins that may well influence the production and stability of uterine ribosomes are ribonucleases and ribonuclease inhibitors. It might be expected, for instance, that cytoplasmic ribonuclease would be less active in hormone-stimulated animals and that any ribonuclease inhibitor would be more active. Such findings would be in line with those of workers studying a number

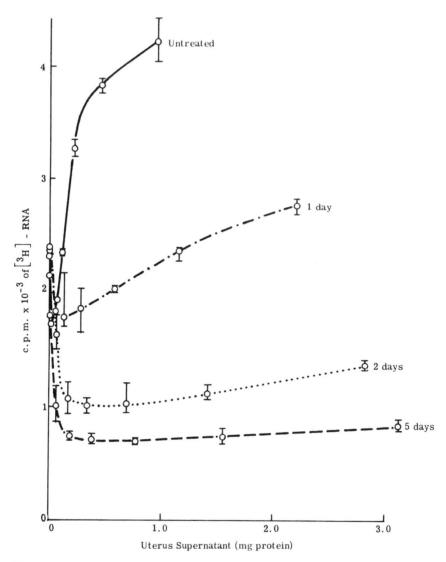

Figure 10.   Effect of estradiol-17β implants on the activity of uterine ribonuclease inhibitor. Radioactivity remaining as acid-precipitable RNA after the incubation of 5000 cpm of [³H]RNA (250 μg) for 15 min in the presence of 0.01 μg of pancreatic ribonuclease and various volumes of uterine cytoplasm. The error bars represent standard deviations derived from duplicate analyses on each of two separate preparations; ———, control rats; — • — • —, rats receiving estrogen implants 1 day before death; • • • •, rats receiving estrogen implants 2 days before death; – – –, rats receiving estrogen implants 5 days before death. (Reproduced from McGregor *et al.* (1981) by courtesy of the Editor-in-Chief of the *Journal of Steroid Biochemistry*.)

of other growth systems (Brewer *et al.*, 1969; Moriyama *et al.*, 1969; Kraft and Shortman, 1970). A ribonuclease inhibitor is present in immature rat uterus at levels roughly comparable with the well-studied liver inhibitor. In contrast to the above systems, however, estrogen treatment results in the total disappearance of the inhibitor from the uterus while having no effect on that of the liver (Fig. 10). The uterine inhibitor also disappears during the normal development of the rat and can not be detected in adults at any point in the estrous cycle (Munro and Knowler, 1981). We are currently attempting to determine whether the disappearance is a transcriptional event. If so, it reveals that estrogen can function in negative as well as positive transcriptional control.

Concomitantly with the loss of inhibitor activity, cytoplasmic ribonuclease activities are stimulated (McGregor *et al.*, 1981; Munro and Knowler, 1981) but we do not at present know whether this can be fully explained by the disappearance of the inhibitor or whether other factors are involved.

6.4. Nonhistone Chromatin Proteins

It seems most likely that the proteins that modulate estrogen-induced rDNA transcription are chromatin proteins, possibly nonhistone proteins specifically associated with the nucleolus. New nonhistone proteins, made in response to estrogen, have been detected on one-dimensional polyacrylamide gels (King *et al.*, 1974; Cohen and Hamilton, 1975) but such systems only detect the most abundant protein species which are likely to have a structural rather than controlling role. The search for minor nonhistone chromatin proteins that may specifically control rRNA transcription will prove very difficult and the techniques of genetic manipulation appear at present to form a probe of limited value. Thus, we have recently cloned, in plasmid pBR322, 71 sequences that encode mRNA species induced in rat uterus between 2 and 4 hr after the administration of estrogen (W. Bajwa and J. T. Knowler, unpublished). These clones are currently being employed in filter hybridizations and "Northern blot" analyses to quantitate hormone-induced changes in the levels of mRNAs for which they are complementary. However, the further screening of these 71 clones (let alone the much larger number necessary to ensure success) for sequences that encode nonhistone chromatins presents considerable problems.

7. Investigations of the Mode of Action of the Antiestrogen, Tamoxifen

The antiestrogen, tamoxifen, is an effective chemotherapeutic agent in the treatment of metastatic mammary carcinoma. Studies on its mode of action have established that it binds to the estrogen cytoplasmic receptor (Jordan and Prestwick, 1977) and that the tamoxifen–receptor complex is translocated to the nucleus

(Nicholson *et al.*, 1977; Koseki *et al.*, 1977). We have been interested in studying nuclear events in the immature rat uterus responding to the administration of tamoxifen or a combination of estrogen and tamoxifen and have shown that the interrelationship between the two compounds is complex.

As previously described, a single administration of estradiol-17β causes a stimulation of uterine RNA synthesis which peaks at 6 to 12 times control levels 2–4 hr after its injection. Tamoxifen, on the other hand, at an optimal dosage of 1–2 mg/kg injected subcutaneously in oil, had no effect at 2 hr but induced a broad peak of stimulated RNA synthesis which lasted from 12 to 32 hr after its administration and peaked at seven times control levels at 24 hr (Waters and Knowler, 1981a). An agonist effect of the inhibitor when administered alone was not unexpected but its time course was. It differed from the results of other workers looking at the effects of tamoxifen on RNA polymerase activity in isolated uterine nuclei (Davies *et al.*, 1979; Kurl and Borthwick, 1980) and it differed from our results with other agonists. Thus, estriol, which in our systems behaves as a weak agonist, elicits a stimulation of RNA synthesis which is much less than that produced by estradiol-17β but nevertheless follows an identical time course (Knowler, 1978). Our more recent results (Waters *et al.*, 1982) suggest the delayed but prolonged agonist action of tamoxifen may be explained, firstly by its long plasma half-life (Adam *et al.*, 1980); secondly by its metabolism, mainly in liver, to more active hydroxylated derivatives; and thirdly by the slow accumulation of the active metabolites in the uterus. Borgna and Rochefort (1981) have shown that 4-hydroxy tamoxifen is a much more effective competitor for the estrogen receptor than tamoxifen and our preliminary findings indicate that it is a much more estrogenlike agonist. Thus, the time course of 4-hydroxy tamoxifen-stimulated uterine RNA synthesis reveals an estradiollike peak 2–4 hr after administration.

The above concepts cannot, however, fully explain the effect on uterine RNA synthesis of combined administrations of estradiol and tamoxifen. Thus, it is not surprising that tamoxifen is unable to totally inhibit estradiol-induced RNA synthesis (Waters and Knowler, 1981a), but we are unable to explain why, in the presence of estrogen, the peak of tamoxifen-stimulated RNA synthesis was partially abolished 12 hr after administration and almost totally abolished 24 hr after administration (Fig. 11). Thus, in some as yet ill-understood way, estrogen appears to be able to inhibit the transcriptional effects of tamoxifen even though the inhibition occurs at a time when one would expect the estradiol to have been metabolized. It may be that the recent observations that tamoxifen associates with "receptor proteins" other than the estradiol receptor (Faye *et al.*, 1981; Nawata *et al.*, 1981) can in part explain this effect.

The above studies were initially performed by monitoring the incorporation of precursor into total uterine RNA (Waters and Knowler, 1981a) but they have recently been extended and confirmed with fractionated RNA (Waters and Knowler, 1981b). Figure 11 shows that all of the antagonistic and agonistic effects of

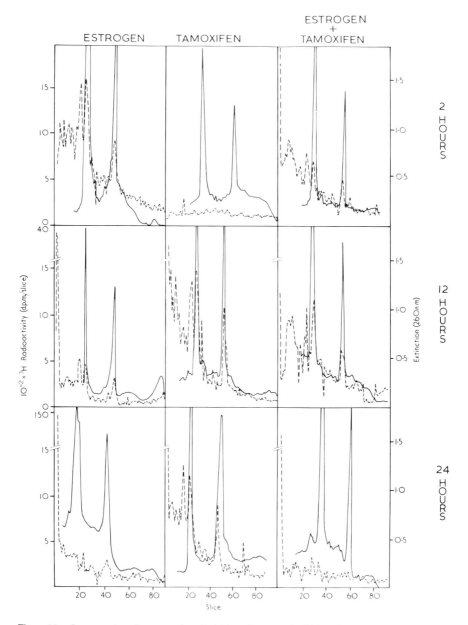

Figure 11.   Incorporation of precursor into the high-mol.-wt. uterine RNA of immature rats responding to estrogen and tamoxifen. Groups of eight rats aged 19–21 days, weighing 30–35 g, received subcutaneous injections of 0.1 ml of corn oil in which was dissolved 1.0 μg of estradiol-17β or 30 μg of tamoxifen or a combination of both at 2, 12, or 24 hr before death. They subsequently received radioactive RNA-precursor 30 min before death and high-mol.-wt. uterine RNA was prepared and analyzed on 2.7% polyacrylamide gels. For reasons of clarity the E260 nm trace is omitted from the beginning of the gels. ———, E260 nm; – – –, ³H radioactivity in dpm/slice. (Reproduced from Waters and Knowler (1981b) by courtesy of the editors of *FEBS Letters*.)

tamoxifen could be observed when high-mol.-wt. uterine RNA species were fractionated on polyacrylamide gels. 28 S and 18 S ribosomal RNA, 45 S and 32 S pre-ribosomal RNA and hnRNA were all involved in the response. It is hoped that further studies on the effect of the inhibitor on estrogen-induced changes in mRNA populations and on estrogen-induced ribosome accumulation will clarify the site of action of the inhibitor and perhaps assist in the selection of cloned cDNA sequences for our continuing study of the mode of action of estrogen itself.

REFERENCES

Adam, H. K., Patterson, J. S., and Kemp, J. V., 1980, Studies on the metabolism and pharmacokinetics of tamoxifen in normal volunteers, *Cancer Treat. Rep.* **64**:761–764.

Aziz, S., and Knowler, J. T., 1978, Characterization of uterine heterogeneous nuclear ribonucleic acid and the effect of estradiol-17β on its synthesis, *Biochem. J.* **172**:587–593.

Aziz, S., and Knowler, J. T., 1980, The detection of mRNA sequences in hnRNA fractions of the estrogen-stimulated rat uterus, *Biochem J.* **187**:265–267.

Aziz, S., Balmain, A., and Knowler, J. T., 1979a, Diversity and complexity of uterine mRNA from rats of differing hormonal states, *Eur. J. Biochem.,* **100**:85–94.

Aziz, S., Balmain, A., and Knowler, J. T., 1979b, Qualitative and quantitative changes in uterine mRNA populations in response to estradiol treatment of rats, *Eur. J. Biochem.* **100**:95–100.

Beaumont, J., and Knowler, J. T., 1982, The effect of estradiol on protein synthesis in the uterus of the immature rat, *Trans. Biochem. Soc.* **10**:50.

Bishop, J. O., Morton, J. G., Rosbach, M., and Richardson, M., 1974, Three abundance classes in HeLa cell mRNA, *Nature* **250**:199–203.

Borgna, J-L., and Rochefort, H., 1981, Hydroxylated metabolites of tamoxifen are formed *in vivo* and bound to estrogen receptor in target tissues, *J. Biol. Chem.* **256**:859–868.

Borthwick, N. M., 1974, The action of estradiol-17β on RNA polymerases in the uterus of the immature rabbit. Ph.D. thesis, Department of Biochemistry, University of Glasgow, Glasgow, Scotland, U.K.

Borthwick, N. M., and Smellie, R. M. S., 1975, The effects of estradiol-17β on the RNA polymerases of immature rabbit uterus, *Biochem. J.* **147**:91–101.

Brewer, E. H., Foster, L. B., and Snells, B. H., 1969, A possible role for ribonuclease in the regulation of protein synthesis in normal and hypophysectomized rats, *J. Biol. Chem.* **244**:1389–1392.

Cohen, M. E., and Hamilton, T. H., 1975, Effect of estradiol-17β on the synthesis of specific uterine non-histone chromosomal proteins. *Proc. Natl. Acad. Sci. USA* **72**:4346–4350.

Davies, P., Syne, J. S., and Nicholson, R. J., 1979, Effects of estradiol and the antiestrogen tamoxifen on steroid hormone receptor concentration and nuclear RNA polymerase activities in rat uteri, *Endocrinology* **105**:1336–1342.

Faye, J-C., Lasserre, B., and Bayard, F., 1980, Antiestrogen specific, high affinity saturable binding sites in rat uterine cytosol, *Biochem. Biophys. Res. Commun.* **93**:1225–1231.

Galau, G. A., Britten, R. J., and Davidson, E. H., 1974, A measurement of the sequence complexity of polysomal mRNA in sea urchin embryos, *Cell* **2**:9–20.

Glasser, S. R., Chytil, F., and Spelsberg, T. C., 1972, Early effects of estradiol-17β on the chromatin and activity of DNA dependent RNA polymerases I and II of the rat uterus, *Biochem. J.* **130**:947–957.

Hadjiolov, A., Dabeva, M., and Mackedonski, V., 1974, The action of α amanitin *in vivo* in the synthesis and maturation of mouse liver ribonucleic acids. *Biochem. J.* **138**:321–334.

Jordan, V. C., and Prestwick, T., 1977, Binding of [³H] tamoxifen in rat uterine cytosols: A comparison of swinging bucket and vertical tube rotor sucrose density gradient analysis. *Mol. Cell Endocrinol.* **8**:179–188.

Kaye, A. M., and Reiss, N., 1980, The uterine "estrogen-induced protein" (IP): purification, distribution and possible function, in: *Steroid Induced Proteins* (M. Beato, ed.), pp 3–19, Elsevier/North Holland Biochemical Press, Amsterdam.

King, R. J. B., Somjen, D., Kaye, A. M., and Lindner, H. R., 1974, Stimulation by estradiol-17β of specific cytoplasmic and chromosomal protein synthesis in immature rat uterus, *Mol. Cell Endocrinol.* **1**:21–36.

Knecht, D. A., and Luck, D. N., 1977, Synthesis and processing of ribosomal RNA by the uterus of ovariectomized adult rat during early estrogen action, *Nature* **266**:563–564.

Knowler, J. T., 1976, The estrogen stimulated synthesis of RNA in nuclear ribonucleoprotein particles of the immature rat uterus, *Eur. J. Biochem.* **64**:161–165.

Knowler, J. T., 1978, Oestriol-stimulated synthesis of ribonucleic acid in the uterus of the immature rat, *Biochem. J.* **170**:181–183.

Knowler, J. T., and Smellie, R. M. S., 1971, The synthesis of ribonucleic acid in immature rat uterus responding to estradiol, *Biochem. J.* **125**:605–614.

Knowler, J. T., and Smellie, R. M. S., 1973, The estrogen-stimulated synthesis of heterogeneous nuclear ribonucleic acid in the uterus of immature rats, *Biochem. J.* **131**:689–697.

Knowler, J. T., Borthwick, N. M., and Smellie, R. M. S., 1975, Early effects of estrogen on the transcription of uterine ribonucleic acid, *Biochem. Soc. Trans.* **3**:1177–1180.

Korach, K. S., and Ford, E. B., 1978, Estrogen action in the mouse uterus: An additional nuclear event. *Biochem. Biophys. Res. Commun.* **83**:327–333.

Koseki, Y., Zava, D. T., Chamness, G. C., and McGuire, W. L., 1977, Estrogen receptor translocation and replenishment by the antiestrogen tamoxifen. *Endocrinology* **101**:1104–1109.

Kraft, N., and Shortman, K., 1970, A suggested control function for the animal tissue ribonuclease-ribonuclease inhibitor system, based on studies of isolated cells and phytohaemogglutinin transformed lymphocytes, *Biochem. Biophys. Acta* **217**:164–175.

Kurl, R. N., and Borthwick, N. M., 1980, Clomiphen and tamoxifen action in the rat uterus. *J. Endocrinol.* **85**:519–524.

Lindell, T. J., O'Malley, A. F., and Puglisi, B., 1978, Proposed role of mRNA in rRNA transcription, *Biochemistry* **17**:1154–1159.

McGregor, C. W., Adams, A., and Knowler, J. T., 1981, Ribonuclease and ribonuclease inhibitor activity in the uteri of immature rats, *J. Steroid Biochem.* **14**:415–419.

Merryweather, M. J., and Knowler, J. T., 1980, The kinetics of the incorporation of newly synthesized RNA and protein into the ribosomes of the uterus of the estrogen-stimulated immature rat, *Biochem. J.* **196**:405–410.

Miller, B. G., and Baggett, B., 1972, Effects of 17β-estradiol on the incorporation of pyrimidine nucleotide precursors into nucleotide pools and RNA in the mouse uterus, *Endocrinology* **90**:645–656.

Moriyama, T., Umeda, T., Nakashima, S., Oura, H., and Tsukada, K., 1969, Studies on the function of ribonuclease inhibitor in rat liver, *J. Biochem.* **66**:151–156.

Munro, J., and Knowler, J. T., 1981, Changing levels of ribonuclease and ribonuclease inhibitor in the uterus of the developing rat, *J. Steroid Biochem.* **16**:293–295.

Nawata, H., Bromzert, D., and Lippman, M. E., 1981, Isolation and characterization of a tamoxifen-resistant cell line derived from MCF-7 human breast cancer cells, *J. Biol. Chem.* **256**:5016–5021.

Nicholson, R. I., Davies, P., and Griffiths, K., 1977, Effects of estradiol-17β receptors in DMBA-induced rat mammary tumors, *Eur. J. Cancer* **13**:201–298.

Nicolette, J. A., and Babler, M., 1974, The selective inhibitory effect of NH₄Cl on Estrogen-stimulated rat uterine *in vitro* RNA synthesis. *Arch. Biochem. Biophys.* **163**:656–665.

Notides, A., and Gorski, J., 1966, Estrogen-induced synthesis of a specific uterine protein, *Proc. Natl. Acad. Sci. USA* **58**:230–235.

O'Farrell, P. H., 1975, High resolution two dimensional electrophoresis of proteins, *J. Biol. Chem.* **250**:4007–4021.

Oliver, J. M., 1971, The effects of estradiol on nucleoside transport in rat uterus, *Biochem. J.* **121**:83–88.

Oliver, J. M., and Kellie, A. E., 1970, The effects of estradiol on the acid-soluble nucleotides of rat uterus, *Biochem. J.* **119**:187–191.

Parker, M. G., and Mainwaring, W. I. P., 1977, Effects of androgens on the mRNA complexity of poly(A) RNA from rat prostate, *Cell* **12**:401–407.

Raynaud-Jammet, C., Catellie, M. G., and Baulieu, E-E., 1972, Inhibition by amanitin of the estradiol-induced increase in α amanitin insensitive RNA polymerase in immature rat uterus, *FEBS Letts.* **22**:93–96.

Reiss, N. A., and Kaye, A. M., 1981, Identification of the major component of the estrogen-induced protein of rat uterus as the BB isoenzyme of creatine kinase, *J. Biol. Chem.* **256**:5741–5749.

Schmid, W., and Sekeris, C. E., 1973, Possible involvement of nuclear DNA-like RNA in the control of ribosomal RNA synthesis, *Biochim. Biophys. Acta.* **312**:549–554.

Waters, A. P., and Knowler, J. T., 1981a, A comparison of the effects of estrogen and tamoxifen on the synthesis of uterine RNA in immature rats. *J. Steroid Biochem.* **14**:625–630.

Waters, A. P., and Knowler, J. T., 1981b, A comparison of the effects of estrogen and tamoxifen on the synthesis of uterine high molecular mass RNA, *FEBS Letts.* **129**:17–20.

Waters, A. P., Wakeling, A. E., and Knowler, J. T., 1983, A rationalization of the effects of anti-oestrogens on uterine RNA synthesis in the immature rat, *J. Steroid Biochem.,* **18**:7–11.

Webster, R. A., and Hamilton, T. H., 1976, Comparative effects of estradiol-17β and oestriol on uterine RNA polymerases I, II and III *in vivo, Biochem. Biophys. Res. Commun.* **69**:737–743.

Wells, D. J., 1973, Partial purification and characterization of RNA polymerase from chick oviduct. Ph.D. Thesis, Department of Obstetrics and Gynecology, Vanderbilt University, Nashville, Tennessee, U.S.A.

Young, B. D., Harrison, P. R., Gilmour, R. S., Birnie, G. D., Hell, A., Humphries, S., and Paul, J., 1974, Kinetic studies of gene frequency, *J. Mol. Biol.* **84**:555–568.

8

Regulation of the Expression of the Uteroglobin Gene by Ovarian Hormones

Miguel Beato, Jutta Arnemann, Carla Menne, Heidrun Müller, Guntram Suske, and Michael Wenz

1. Introduction

The elucidation of the molecular mechanisms by which eukaryotic cells modulate the expression of specific genes remains one of the major challenges for modern biochemistry. Since the hypothesis was put forward that steroid hormones act as gene regulators (Karlson, 1961), they have been widely used as inducers in model experimental systems in the hope that clarification of their mechanism of action will help us to understand the process of regulated gene expression. Despite considerable progress, however, the molecular mechanisms underlying hormone control of gene expression remain elusive. Most of what we know at the molecular level comes from avian systems such as the induction of egg-white proteins in the oviduct or vitellogenin in the liver. In these systems both the hormone receptors and the regulated genes have been carefully characterized, but the mechanisms of their interaction are not understood.

The discovery of uteroglobin as a major protein component of the uterine

Miguel Beato, Jutta Arnemann, Carla Menne, Heidrun Müller, Guntram Suske, and Michael Wenz ● Institut für Physiologische Chemie der Phillips-Universität, 3550 Marburg, G.F.R.

secretion of pregnant rabbits (Krishnan and Daniel, 1967; Beier, 1968), and the demonstration that its secretion is dependent on ovarian hormones (Beier, 1968; Arthur and Daniel, 1972) offered the possibility of exploring a mammalian induction system, bearing some of the advantages of the avian system. Until that time, the most popular mammalian systems had been the hormonal induction of enzymes, such as tyrosine aminotransferase, and no major secretory or structural protein was known to be under the control of steroid hormones.

2. Structure and Function of Uteroglobin

Uteroglobin has now been purified to homogeneity (Nieto et al., 1977), and its primary and quaternary structures have been elucidated (Ponstingl et al., 1978; Popp et al., 1978; Atger et al., 1979). In agreement with previous reports (Murray et al., 1972; McGaughey and Murray, 1972) it was found that uteroglobin is composed of two identical polypeptide chains, each 70 amino acids long, held together by two disulfide bonds (Nieto et al., 1977; Ponstingl et al., 1978). Several crystal forms of the native oxidized protein have been obtained (Buehner and Beato, 1978; Mornon et al., 1978, 1979), and recently the structure of an oxidized form of the protein containing antiparallel-oriented polypeptide chains has been established at a resolution of 2.2 Å (Mornon et al., 1980).

In spite of our detailed knowledge of uteroglobin structure virtually nothing is known about its physiological function. The initial proposition, that uteroglobin, also called blastokinin, could be required for the development of the preimplantation embryo (Krishnan and Daniel, 1967; Gulyas et al., 1969; El-Banna and Daniel, 1972a,b), has not been substantiated (Maurer and Beier, 1976). The idea that uteroglobin acts on blastocyst development by inhibiting protease activity (Beier et al., 1980) has also been abandoned, as highly purified preparations of uteroglobin do not exhibit such activity (H. M. Beier, personal communication). Another interesting but merely speculative hypothesis is that uteroglobin acts as an immunosuppressor by cross-linking to the β_2-microglobulin at the surface of embryonic cells (Mukherjee et al., 1980). This suggestion is based on a comparison of the amino acid sequence of both proteins and on the observation that transglutaminase activity in rabbit endometrium is progesterone-dependent. However, no experimental evidence has been provided for the existence of cross-linked uteroglobin.

The only well-established activity of uteroglobin is its ability to bind progesterone and related steroids. The initial reports on this activity (Urzua et al., 1970; Arthur et al., 1972) were later questioned (Goswami and Feigelson, 1974; Rahman et al., 1975b), but have been confirmed with highly purified uteroglobin preparations and precise binding experiments (Beato and Baier, 1975; Beato, 1976; Fridlansky and Milgrom, 1976; Beato et al., 1977). Reduction of the disulfide bonds of "native" uteroglobin is required for efficient progesterone binding, but the sulfhydryl groups are not directly implicated in the interaction with the steroid (Beato and Baier, 1975; Beato *et al.*, 1977). Although the protein remains as a dimer after reduction, considerable structural rearrangements are observed, as has been shown by proton magnetic resonance techniques (Puigdoménech and Beato, 1977). Contrary to a recent proposal (Mornon *et al.*, 1980), we do not find a molecule of progesterone bound to the "native" oxidized form of uteroglobin (Tancredi *et al.*, 1982). Nuclear magnetic resonance (NMR) techniques and studies on the influence of chemical modifications of amino acid side chains of uteroglobin on progesterone binding have provided considerable information on the structural aspects of the protein relevant for the interaction with the steroid (Saavedra and Beato, 1980; Temussi *et al.*, 1980). A detailed revision of these data has been published recently (Beato *et al.*, 1980a).

The progesterone-binding capacity of uteroglobin has generated further hypothesis regarding its functional role. In principle the protein could act as a steroid carrier from the maternal organism to the blastocyst, or as progesterone scavenger (Beato, 1977b) that protects the early embryo from damage by exogenous progesterone (Whitten, 1957; Daniel, 1964; Daniel and Levy, 1964; Roblero, 1973). Recently, it has been suggested that uteroglobin could serve to maintain a high concentration of progesterone in the uterine lumen, thus allowing a more intense progestational transformation of the preimplantation endometrium (Bochskanl and Kirchner, 1981). This and other hypotheses based on steroid-binding properties of uteroglobin, should take into consideration the fact that a uteroglobinlike protein has been found in organs other than the uterus (see below).

3. Distribution of Uteroglobin and Differential Hormonal Control

Uteroglobin has so far been conclusively found only in the rabbit. Although the uterine secretions of other species contain uterine-specific proteins none of these proteins exhibit the characteristic properties of uteroglobin. The presence of uteroglobin in human uterine secretion has been the subject of contradictory reports (Beier *et al.*, 1971; Bernstein *et al.*, 1971; Daniel, 1971, 1972; Shirai *et al.*, 1972; Wolf and Mastroianni, 1975). Using highly specific antibodies and progesterone-binding techniques, we could not detect uteroglobinlike proteins in human uterine fluids collected all through the menstrual cycle (Voss and Beato, 1977).

In the rabbit, uteroglobin has not only been found in the uterus, but also in the oviduct, the male genital tract, the digestive tract, and the respiratory tract of both males and females (Kay and Feigelson, 1972; Petzoldt *et al.*, 1972; Goswami and Feigelson, 1974; Beier *et al.*, 1975; Daniel and Milazzo, 1976; Kirchner, 1976a; Kirchner and Schroer, 1976; Noske and Feigelson, 1976). Interestingly, the hormonal regulation of uteroglobin secretion differs in the different tissues. Whereas in the endometrium progesterone, and to a lesser extent estradiol, are effective inducers (see below), in the oviduct, for instance, estradiol is the only inducing steroid (Goswami and Feigelson, 1974). It has been claimed that the lack of progesterone effect on uteroglobin secretion in the oviduct could be due to the absence of the corresponding hormone receptor (El-Banna and Sacher, 1977), but this does not seem a valid explanation for other tissues. In the lung, a protein very similar to uteroglobin is found both in male and female rabbits. This protein exhibits progesterone-binding properties identical to uterine uteroglobin (Beato and Beier, 1978), but it is not induced by progesterone (Feigelson *et al.*, 1977; Torkkeli *et al.*, 1978), although a progesterone receptor is detectable in lung tissue (Savouret *et al.*, 1980). In this organ, the synthetic glucocorticoid dexamethasone is able to increase by three times the synthesis of uteroglobin (Torkkeli *et al.*, 1978; Savouret *et al.*, 1980).

4. Hormonal Regulation of Uteroglobin Synthesis: Studies with Isolated Uteri, Endometrial Explants, and Endometrial Cell Cultures

Uteroglobin is detectable in small amounts in the uterine fluid of estrous rabbits (Kirchner, 1972; Bullock and Connell, 1973; Mayol and Longenecker, 1974), and can be induced in castrated rabbits by the administration of progesterone (Arthur and Daniel, 1972). Optimal induction, however, requires the sequential treatment of the animal with estradiol and progesterone (Bullock and Willen, 1974; Mayol and Longenecker, 1974; Beato and Arnemann, 1975; Rahman *et al.* 1975a). At the time of maximal induction, in rabbits made pseudopregnant by sequential treatment with estradiol and progesterone, uteroglobin accounts for up to 50% of the total protein content of the uterine secretion. These findings are based on quantitative analysis of stained polyacrylamide gels, and do not yield information on the site or the rate of synthesis of the protein. Using immunofluorescence techniques Johnson (1972) and Kirchner (1972) initially demonstrated the accumulation of uteroglobin in the epithelial cells of the endometrium. Later these studies have been refined and extended to other tissues and secretory proteins other than uteroglobin (Kirchner, 1976b, 1979, 1980). But still this type of experiment does not allow the conclusion that uteroglobin is synthesized in the endometrium. This conclusion was made very probable by the observation of Krishnan (1970), Daniel (1971), and Murray and Daniel (1973) that radioactive amino acids injected into the uterine lumen are preferentially incorporated into uteroglobinlike fractions.

Using the technique of intraluminal perfusion and a specific antiuteroglobin antiserum, we showed that both estradiol and progesterone do actually increase the rate of synthesis of uteroglobin and not only the extent of its secretion (Beato and Arnemann, 1975). In fact, there was a good correlation between uteroglobin content in the uterine lumen before perfusion, and the *in vitro* incorporation of radioactive amino acids into uteroglobin. Later these experiments have been confirmed and extended using endometrial explants cultured *in vitro* (Joshi and Ebert, 1976; Nieto and Beato, 1980). In these experiments, we measured uteroglobin levels by radioimmunoassay and the rate of synthesis of uteroglobin by the incorporation of [^{35}S]methionine. We found that estradiol has a more pronounced effect on uteroglobin synthesis than on uteroglobin secretion, whereas progesterone increases both parameters to a much greater extent. Under optimal hormonal conditions, that is in pseudopregnant animals, uteroglobin accounts for up to 9% of the total protein synthesized in the endometrium and for 50–60% of the secreted proteins (Table I) (Nieto and Beato, 1980). In addition we showed that the half life of uteroglobin is not influenced by hormonal treatment and that there is no accumulation of a uteroglobin precursor in the tissue.

In interpreting these results one should keep in mind that the endometrium is a heterogeneous tissue, in which the epithelial cells that synthesize uteroglobin represent only a fraction of the cell population, varying from about 20% in the control estrous rabbits to 30–40% in estradiol-treated animals, and about 50% in pseudopregnant rabbits (O'Toole and H.M. Beier, personal communication). The difficulties due to the heterogeneous nature of the endometrium would be avoided if a cell culture of endometrial epithelial cells were available. There are several reports in the literature concerning the cultivation of endometrial cells from rabbit (Gerschenson *et al.*, 1974, 1977; Whitson and Murray, 1974; Berliner and Gerschenson, 1976; Gerschenson and Berliner, 1976; Murai *et al.*, 1978), rat (Sonnenschein *et al.*, 1974), and humans (Chen *et al.*, 1973; Pavlik and Katzenellenbogen, 1978). Of these reports only one describes the influence of

Table I. Influence of Hormonal Treatment on Several Parameters on the Pathway to Uteroglobin Synthesis

| Treatment | UG-synthesis[a] | | UG-mRNA content[b] | | | | UG-mRNA synthesis[c] | |
| | | | Total | | Nuclear | | | |
	%	Effect	%	Effect	%	Effect	%	Effect
Control	0.67	(1)	0.0014	(1)	0.0026	(1)	0.0039	(1)
Estradiol[d]	1.86	(2.8)	0.0056	(4.0)	0.0131	(5.0)	0.0125	(3.2)
Pseudopregnant[e]	9.14	(13.6)	0.0172	(12.3)	0.0387	(14.9)	0.0485	(12.4)
p[f]		4.9		3.1		3.0		3.9

[a] Measured in endometrial explants (Nieto and Beato, 1980).
[b] Measured by cDNA excess hybridization (Mller and Beato, 1980; Heins and Beato, 1981).
[c] Measured in a nuclear transcription system (Mller and Beato, 1980).
[d] Three injections (100 μg each) on days 1, 3, and 5. Animals were killed on day 7.
[e] Two daily injections of estradiol (100 μg each) followed by five daily injections of progesterone (2 mg each). Animals were killed 24 hr after last injection.
[f] Ratio between the values found in pseudopregnant and estradiol-treated animals.

ovarian hormones on uteroglobin synthesis (Whitson and Murray, 1974), claiming that uteroglobin is induced by progesterone but inhibited by estrogens. These results, however, could not be reproduced, and they probably represent an artifact due to the use of fetal calf serum in the culture system (Hossner, 1976). In conclusion, there is at present no experimental system in which the induction of uteroglobin by ovarian hormones could be studied *in vitro*. Therefore, such important questions as the exact kinetics of induction and the influence of inhibitors of protein synthesis on the induction mechanism cannot be answered precisely.

5. Characterization of Uteroglobin mRNA and Cell-Free Translation Experiments

The increased role of synthesis of uteroglobin in the endometrium of hormonally treated rabbits could be due to an accumulation of uteroglobin mRNA or to a more efficient translation of preexisting mRNA. To answer this question it was necessary to characterize and quantitate the amount of uteroglobin mRNA present in endometrial cells under different hormonal conditions. After a successful translation of uteroglobin mRNA in *Xenopus* oocytes (Beato and Runger, 1975) we decided to use cell-free translation systems for mRNA titration experiments. Both in a Krebs II ascites system (Schütz *et al.*, 1974) and in a wheat germ system (Roberts and Paterson, 1973), the polyadenylated RNA from endometrial polysomes of pseudopregnant rabbits was translated into a precursor of the uteroglobin monomer, preuteroglobin, that was characterized immunologically and by tryptic digestion, and was shown to contain some 20 additional amino acids at its N-terminal end (Beato and Nieto, 1976). These results have been confirmed (Bullock *et al.*, 1976; Levey and Daniel, 1976; Atger and Milgrom, 1977; Bullock, 1977a,b), and the sequence of the 21 amino acids of the signal peptide at the N-terminal end of preuteroglobin has been established (Atger *et al.*, 1979; Malsky *et al.*, 1979). Bullock (1977a) also showed that preuteroglobin mRNA was present in polyadenylated RNA from rabbit lung, and these data have also been confirmed later (Savouret *et al.*, 1980; Heins and Beato, 1981). Using the *in vitro* translation assays we have shown that there is very little uteroglobin mRNA in the endometrial polysomes of estrous rabbits, and that following administration of estradiol the level of uteroglobin mRNA in polysomes can be increased by three times (Beato and Nieto, 1976). After subsequent administration of progesterone there is a 17-fold increase in the polysomal level of uteroglobin mRNA (Beato and Nieto, 1976), in good agreement with the measurements of uteroglobin rate of synthesis mentioned above (Nieto and Beato, 1980). Similar results have been obtained with normally pregnant animals (Bullock *et al.*, 1976; Atger and Milgrom, 1977). Thus, the increased rate of uteroglobin synthesis observed in hormonally treated animals is mainly due to an accumulation of uteroglobin mRNA in polysomes.

The quantitation of mRNA in cell-free translation systems suffers from two main limitations. Firstly, only translatable polyadenylated mRNA molecules are measured, and secondly the efficiency of translation is not equal for all mRNA. In our experiments there is a preferential translation of uteroglobin mRNA, that is probably due to its small size, and this leads to an overestimation of its cellular levels (Beato and Nieto, 1976; Savouret *et al.*, 1980). A more precise titration of uteroglobin mRNA sequences can be achieved by the use of molecular hybridization to a cDNA probe. The preparation of this cDNA probe requires the purification of uteroglobin mRNA.

6. Titration of Uteroglobin mRNA by Molecular Hybridization

Initial attempts to purify uteroglobin mRNA by immunological techniques were unsuccessful as the antibodies raised against the mature protein do not bind efficiently to nascent preuteroglobin chains in polysomes (Beato, 1977a). The uteroglobin mRNA was therefore purified from endometrial polysomes of pseudopregnant rabbits by a combination of oligo(dT)-cellulose chromatography, sucrose density gradient centrifugation, and polyacrylamide gel electrophoresis (Arnemann *et al.*, 1977). Uteroglobin mRNA isolated under these conditions has a sedimentation coefficient of 9 S and an apparent mol. wt. of 200,000 in polyacrylamide gels containing 99% formamide, corresponding to some 600 nucleotides. The 5'-end of the mRNA is capped and the poly(A)-tail at the 3'-end is about 60 nucleotides long (Arnemann *et al.*, 1977). Since 273 nucleotides will be sufficient to encode the 91 amino acid residues of preuteroglobin, nearly half the nucleotides in uteroglobin mRNA are not translated. Similar results were obtained by Atger and Milgrom (1977) using endometrium from 5-day pregnant rabbits.

More than 90% of the polypeptides synthesized in cell-free systems supplemented with partially purified uteroglobin mRNA are immunologically and electrophoretically identical to preuteroglobin (Arnemann *et al.*, 1977). Nevertheless, when this mRNA is used as a template for cDNA synthesis, a heterogeneous population of products is obtained, containing less than 50% sequences complementary to uteroglobin mRNA (Arnemann *et al.*, 1979). If, instead of total polysomes, membrane-bound polysomes are used as starting material, better preparations of mRNA are obtained, containing up to 70% uteroglobin mRNA (Arnemann *et al.*, 1979). A cDNA preparation made with this mRNA was used to purify uteroglobin cDNA by back hybridization to the template at low r_0t values, followed by reisolation of the hybrids. This procedure yields a cDNA that is more than 90% homogeneous as judged by hybridization kinetics, and represents a faithful copy of uteroglobin mRNA (Arnemann *et al.*, 1979). Using this cDNA and RNA-excess hybridization techniques we have shown that the endometrium of pseudopregnant rabbits contains considerably more uteroglobin mRNA sequences than the endometrium of control estrous rabbits (Arnemann *et al.*, 1979).

A more accurate titration of uteroglobin mRNA levels was carried out by the method of hybridization with cDNA in excess (Heins and Beato, 1981). It was shown that uteroglobin mRNA represents 0.0014% of the total RNA in control endometrium. This value rises to 0.005% after treatment with estradiol, and to 0.017% after subsequent treatment with progesterone (Heins and Beato, 1981). In the lung, uteroglobin mRNA represents between 0.004 and 0.005% of the total RNA in both control and pseudopregnant rabbits (Heins and Beato, 1981). Even after administration of dexamethasone, which results in a threefold increase in uteroglobin content in the lung, there is no measurable change in uteroglobin mRNA content (Savouret *et al.*, 1980).

The interaction between estradiol and progesterone with respect to uteroglobin gene expression is rather complex. On the one hand, estradiol facilitates optimal progesterone action by inducing the synthesis of the progesterone receptor (Rao and Katz, 1977; Isomaa et al., 1979a), and very low doses of estradiol act synergistically with progesterone (Kopu *et al.*, 1979). On the other hand estradiol, when administered together with progesterone, is able to inhibit the progesterone-induced synthesis of uteroglobin (Kopu *et al.*, 1979; Hemminki *et al.*, 1980). We have found that pretreatment of estrous rabbits with estradiol can lead to an apparent lag phase in their response to progesterone. If the interval between estradiol pretreatment and progesterone administration is long enough (more than 5 days), the response to progesterone, in terms of uteroglobin mRNA accumulation, is very rapid without any indication of a lag phase (Heins and Beato, 1981; Loosfelt *et al.*, 1981).

There is also a difference between the relative effects of estradiol and progesterone on uteroglobin content in uterine secretion and uteroglobin mRNA accumulation in endometrium. Whereas progesterone causes a similar stimulation of both parameters, estradiol, even after prolonged administration, has a more pronounced effect on accumulation of uteroglobin mRNA than on uteroglobin content (Heins and Beato, 1981; Loosfelt *et al.*, 1981). It seems, therefore, that in addition to its effect on the accumulation of uteroglobin mRNA, progesterone has an influence on the mechanism of secretion of uteroglobin. Alternatively estradiol could inhibit efficient uteroglobin secretion.

The antiprogestational activity of estradiol could also account for the cessation of uteroglobin synthesis observed immediately after implantation (Beier *et al.*, 1971; Bullock and Willen, 1974; Barfield *et al.*, 1976; Kopu *et al.*, 1979; Rahman *et al.*, 1981). There is indeed a surge of estrogens at the time of implantation

(Hilliard and Eaton, 1971) and estradiol has been reported to accelerate the normal decay in the levels of uteroglobin and uteroglobin mRNA that follows interruption of progesterone treatment (Kopu *et al.*, 1979). We have confirmed this effect of estradiol on uteroglobin levels, but we have found no effect of estradiol on the levels of uteroglobin mRNA (Heins and Beato, 1981). These data show that administration of estradiol after cessation of progesterone treatment results in the same disparity between mRNA levels and uteroglobin secretion that is observed when estradiol is injected into control animals (see above).

7. Involvement of the Progesterone Receptor

As in other hormone-responsive systems, induction of uteroglobin by progesterone in uterine cells is mediated by a specific receptor. A progesterone-binding protein with the characteristics of a hormonal receptor has been found in rabbit uterus (Faber *et al.*, 1972, 1973; McGuire and Bariso, 1972; Rao *et al.*, 1973; McGuire *et al.*, 1974; Philibert and Raynaud, 1974; Philibert *et al.*, 1977; Tamaya *et al.*, 1977). Using a synthetic progestin, 16α-ethyl-21-hydroxy-19-norpregn-4-ene-3,20-dione, we have carefully analyzed the interaction of different steroids with the uterine progesterone receptor (Fleischmann and Beato, 1978) and have defined the conditions required for the transformation of the receptor into a DNA-binding form *in vitro* (Fleischmann and Beato, 1979). Recently, the receptor has been partially purified (Westphal *et al.*, 1978; Logeat *et al.*, 1981), and we have succeeded in photoaffinity-labeling the progesterone receptor with synthetic progestins (Westphal *et al.*, 1981).

The correlation between receptor content and uteroglobin induction has been intensively studied. Administration of a single dose of progesterone to estrogen-primed animals leads to a rapid and long-lasting decline in the cytosol and total cellular content of progesterone receptor and to a transient nuclear receptor accumulation (Isomaa *et al.*, 1979b; Torkkeli, 1980). A chronic progesterone treatment does not markedly change the cytosol receptor levels but leads to a marked reduction of nuclear receptor content in discrepancy with the classic models of steroid action (Isomaa *et al.*, 1979b; Isotalo *et al.*, 1981). We have confirmed this influence of chronic administration of progesterone on the nuclear levels of its own receptor (Neulen *et al.*, 1982). A similar discrepancy has been observed during normal pregnancy between high preuteroglobin mRNA content and low nuclear progesterone levels (Young *et al.*, 1980). From these studies, a picture is emerging according to which consumption (or processing) of the nuclear progesterone receptor may be an important intermediate step between receptor translocation and the onset of biological response.

8. Hormonal Regulation of Transcription of the Uteroglobin Gene

Since no marked influence of the hormonal treatment on the stability of uteroglobin mRNA has been detected (Heins and Beato, 1981), the increased levels of uteroglobin mRNA can only result from a hormonal regulation of the nuclear synthesis and/or processing of the corresponding precursor RNA. Jänne and colleagues have found a stimulating effect of progesterone on the RNA-synthesizing machinery of rabbit uterus (Kokko *et al.*, 1977; Orava *et al.*, 1979), but no information on transcription of the uteroglobin gene has been available until recently.

Using a nuclear elongation system, prepared from rabbit endometrium, and molecular hybridization techniques, we have been able to show that both estradiol and progesterone influence the rate of transcription of the uteroglobin gene (Müller and Beato, 1980). RNA synthesized in this nuclear system is mercurated and can therefore be separated from endogenous DNA by affinity chromatography on sulfhydryl-Sepharose. This purified RNA was used for titration of uteroglobin mRNA sequences by hybridization to an excess of purified uteroglobin cDNA (Fig. 1). In the liver, an organ where no uteroglobin has been detected, there is virtually no transcription of the uteroglobin gene, whereas in the endometrium of control estrous rabbits uteroglobin mRNA sequences represent 0.0039% of newly synthesized RNA. Thus, there is a basal expression of the gene in control endometrium. Since the hormonal status of these rabbits is not well defined, it would be interesting to measure the level of expression of the gene in long-term ovariectomized animals. Due to the atrophic involution of the uterus, however, not enough endometrial tissue could be collected to perform the necessary titration experiments.

If control estrous rabbits are injected with estradiol for 1 week, one detects a threefold increase in the expression of the uteroglobin gene (Fig. 1; Table I). This value can be further increased, about fourfold, by subsequent injection of progesterone (Fig. 1). Under these conditions of pseudopregnancy, uteroglobin mRNA sequences represent nearly 0.05% of the RNA synthesized in isolated endometrial nuclei, a 12-fold increase over the controls (Table I). The synthesis of uteroglobin RNA sequences can be up to 80% inhibited by addition to the nuclear system of actinomycin D or α-amanitin, thus showing that we measure DNA-dependent RNA synthesis by RNA polymerase II or B (Müller and Beato, 1980).

A comparison of these values with the uteroglobin mRNA content in nuclear RNA shows a good correlation, with the possible exception of the estradiol-treated group (Table I). It seems that the effect of estradiol on the transcription of the uteroglobin gene is slightly less pronounced than its effect on the accumulation of uteroglobin mRNA in nuclear RNA. Although this difference is small and should be statistically substantiated, it could indicate a hormonal

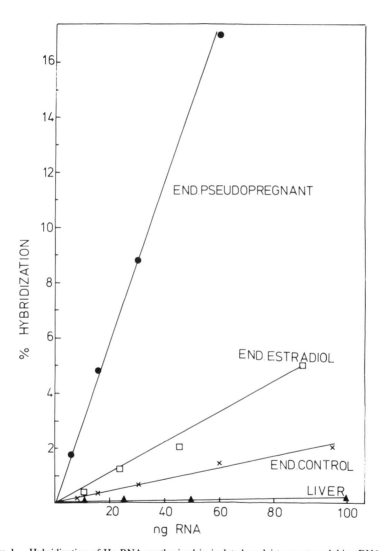

Figure 1. Hybridization of Hg-RNA synthesized in isolated nuclei to preuteroglobin cDNA. Different amounts of each RNA were hybridized to an excess of preuteroglobin [^{32}P]-cDNA at 68° C for 48 hr, and the percentage of cDNA hybridized was determined by digestion with nuclease S1 (Müller and Beato, 1980). The background observed in the absence of added RNA has been subtracted.

influence on the processing of the precursor to uteroglobin mRNA. In addition, Table I shows that in pseudopregnant animals there is a good correlation between uteroglobin synthesis, the cellular and nuclear levels of uteroglobin mRNA, and the transcription rate of the uteroglobin gene. These findings demonstrate that the hormonal regulation of the expression of the uteroglobin gene operates mainly at the level of transcription, and emphasize the need for a better understanding of the fundamental process of specific gene transcription. A prerequisite for studying regulated transcription is the characterization of the transcriptional unit of the uteroglobin gene. In the following two sections we will report our initial attempts to reach this goal.

9. Cloning and Characterization of Uteroglobin cDNA

In order to isolate the uteroglobin gene one needs a homogeneous probe for screening the genomic DNA. Starting with a partially purified uteroglobin mRNA (Arnemann et al., 1979), we synthesized a double-stranded cDNA, that was inserted at the Pst1 site of the cloning vehicle pBR322 by the GC-homopolymer tailing procedure (Menne et al., 1982). After amplification in E. coli χ 1776 (Norgard et al., 1978), clones containing recombinant plasmids were identified by in situ colony hybridization with [³²P]-cDNA (Grunstein and Hogness, 1975), and their plasmid DNA was analyzed by digestion with restriction enzymes. Out of 30 positive clones, 19 had inserts smaller than 300 bp, 10 clones had inserts of around 400 bp, and 1 clone, pcUG-H6, had an insert 520 bp long. This last insert has been completely sequenced, and does not correspond to preuteroglobin (J. Arnemann and M. Beato, unpublished). From the 10 clones containing inserts of around 400-bp length one was similar to pcUG-H6 and the other 9 belong to the same family according to their restriction maps. One of these, pcUG-G8, has been sequenced, by the procedure of Maxam and Gilbert (1980), and its structure is shown in Fig. 2. This insert contains 314 nucleotides of uteroglobin cDNA plus a track of As, corresponding to the poly(A)-tail of the mRNA, and the G- and C-tails incorporated during the cloning procedure. The region corresponding to uteroglobin mRNA includes the complete 3'-untranslated region and the codons for amino acids 13–70 of mature uteroglobin. None of the clones studied so far extends further into the 5'-region of the mRNA, suggesting that structural or other reasons are responsible for the fact that the last one of the 5'-end was not detected (see below).

The nucleotide sequence of pcUG-G8 confirms the amino acid sequence of uteroglobin previously reported (Ponstingl et al., 1978), with the exception of position 61, where we reported Gln, but the nucleotide sequence indicates Glu, in agreement with another report (Popp et al., 1978). Between amino acids 29 and 50 there were eight discrepancies in the reported protein sequences. In all these cases the nucleotide sequence confirms the amino acid sequence reported by Ponstingl *et al.* (1978). The nucleotide sequence for the part of the translated region of uteroglobin cDNA between amino acids 18 and 56 has been reported (Chandra *et al.*, 1980), and it is essentially confirmed by the sequence presented here. Another preliminary report, on the nucleotide sequence corresponding to the amino acids residues 32 to 51 (Atger *et al.*, 1980), shows four conservative and one nonconservative differences when compared to the other two reports (Fig. 2) (Chandra *et al.*, 1980).

Even if the cloned cDNA does not include the complete sequence of utero-globin mRNA it could be used to establish the uteroglobin gene copy number and to isolate the uteroglobin gene from a genomic library.

10. Quantitation, Isolation, and Structural Analysis of the Uteroglobin Gene

To establish the number of copies of the uteroglobin gene per haploid genome a procedure developed by Mirault *et al.* (1979) for the titration of heat–shock genes was used. Different amounts of DNA from the recombinant plasmid pcUG-G8, corresponding to increasing number of gene copies per genome equivalent, were mixed with rabbit endometrium DNA, and hybridized to trace amounts of uteroglobin cDNA labeled with ^{32}P by nick translation (Rigby *et al.*, 1977). With this procedure a standard straight line is obtained from which the number of uteroglobin gene copies per haploid genome can be directly measured (Fig. 3). The results show clearly that the uteroglobin gene is a unique gene present only in one to three copies per haploid genome. Experiments carried out under less stringent hybridization conditions show no indication for the existence of uteroglobin-gene-related sequences in the rabbit genome. Similar results were obtained when liver DNA was used instead of endometrial DNA.

With this information we started the isolation of the uteroglobin gene by screening the rabbit gene library constructed by Maniatis *et al.* (1978) with the plaque hybridization technique of Benton and Davis (1977), using labeled pcUG-G8 DNA as probe. Out of 1×10^6 phage plaques tested, three positive recombinant phages were found and isolated after three purification steps. The inserts in these three phages overlap, and cover a region extending over 35 kb of genomic DNA, that contains uteroglobin gene sequences at its center. A partial restriction map of the three inserts is shown in Fig. 4a, and a 10-fold enlarged map of the region containing uteroglobin gene sequences is shown in Figure 4b. As can be seen, phages λ UG 9.3 and λ UG 1.1 have inserts covering essentially the same genomic region, and containing only the right half of the gene, whereas phage λ UG 9.2 contains the whole uteroglobin gene and extends some 15 kb upwards. Similar results have been recently reported (Atger *et al.*, 1981).

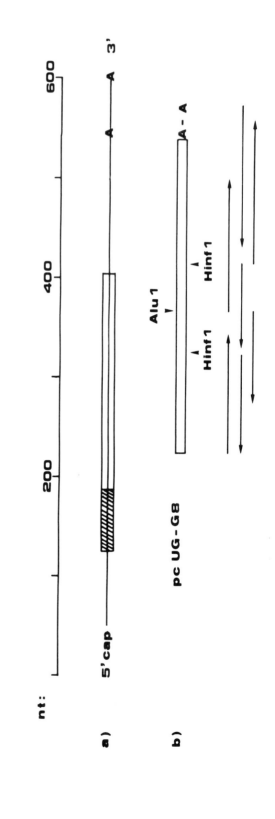

nt:

600

400

200

a)

5' cap

A A   3'

b)

pc UG-G8

A - A

Alu 1

Hinf 1

Hinf 1

c)

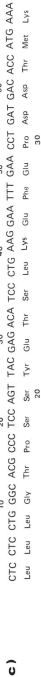

```
          10          20          30               40          50               60
CTC CTC CTG GGC ACG CCC TCC AGT TAC GAG ACA TCC CTG AAG GAA TTT GAA CCT GAT GAC ACC ATG AAA
Leu Leu Leu Gly Thr Pro Ser Ser Tyr Glu Thr Ser Leu Lys Glu Phe Glu Pro Asp Asp Thr Met Lys
                        20                      30

70              90                      110         120                 130
GAT GCA GGG ATG CAG ATG AAG AAG GTG TTG GAC TCC CTG CCC CAG ACG ACC AGA GAG AAC ATC ATG AAG
Asp Ala Gly Met Gln Met Lys Lys Val Leu Asp Ser Leu Pro Gln Thr Thr Arg Glu Asn Ile Met Lys
            40                          50

140             150         160         170         180          190         200          210
CTC ACG GAA AAA ATA GTG AAG AGC CCA CTG TGT ATS TAG GATGGAGGAATCCGAGGTCCTGCGGTCGAGAAGCCGAAG
Leu Thr Glu Lys Ile Val Lys Ser Pro Leu Cys Met end
            60                              70

220         230         240         250         260         270         280         290         300
ATTTCCACCTGCTGAAGCCCCTGCTGCTGCCCCTGGCCCTTGGGTCCCCACCCACCCAACCCAGCCAGCCTTTGCTTTCAATAAACTGCA

310
AGCAGATC (A)₃₀
```

Figure 2. Structure and nucleotide sequence of pcUG-G8. (a) Structure of uteroglobin mRNA as deduced from sizing techniques and DNA sequencing experiments. (b) Structure of pcUG-G8 and restriction sites used for sequencing. The sequence strategy is also shown. (c) Nucleotide sequence of pcUG-G8, aligned with the corresponding amino acid sequence of uteroglobin.

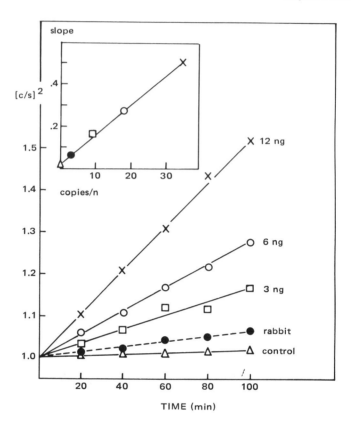

Figure 3.   Determination of uteroglobin gene copy number. High-mol.-wt DNA prepared from the endometrium of induced rabbits was mixed with cloned cDNA labeled by nick translation (specific activity 1.5×10^8 dpm/μg), and with the indicated amounts of pCUG-G8 DNA. After treatment with NaOH to reduce the size to about 600 nucleotides, the DNA was ethanol-precipitated and redissolved in hybridization buffer (Mirault *et al.*, 1979). Poly(A) and poly(C) were added, and the samples were denatured and incubated at 70° C. At the indicated times aliquots were taken for the determination of radioactivity resistant to nuclease S1 digestion. The graphic representation of the data has been described (Mirault *et al.*, 1979). c, total cDNA; s, single stranded cDNA. The "control" corresponds to the probe with labeled DNA alone, and the line labeled "rabbit" represents the sample without added pCUG-G8 DNA. The inset represents a plot of the slope of the individual lines vs. the number of copies of pCUG-G8 DNA per haploid genomic equivalent.

The regions containing uteroglobin gene sequences have been subcloned in the cloning vehicle pBR 322, and sequenced. The results can be summarized as follows:

1. The 5'-end of the gene has been located by S1-nuclease mapping (Suske and Beato, unpublished).
2. The expressed part of the gene, including the 5'-non-translated region, the complete coding region for preuteroglobin, and the complete 3'-nontranslated region is included in a 3-kb DNA fragment.

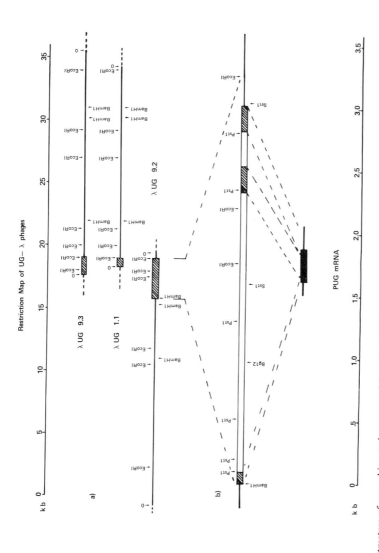

Figure 4. Structure of recombinant phages containing uteroglobin gene sequences. (a) Restriction sites for EcoR1 and BamH1 in the three recombinant phages, λUG 9.3, λUG 1.1, and λUG 9.2. "0" indicates the synthetic EcoR1 sites introduced during the construction of the gene library (Maniatis *et al.*, 1978) and marks the limits of the inserts. The shadowed boxes denotes the presence of uteroglobin sequences. The 35–kb scale on the top is the reference scale for this part of the diagram. (b) Tenfold magnified map of the region containing uteroglobin gene sequences. The shaded regions represent exons and the white regions introns. The PUG mRNA is shown below for comparison. The lower 3.5–kb scale applies to this part of the diagram.

3. The gene is organized into three exons of similar length and two introns of very different length.

4. The first exon contains 47 nucleotides of the non-translated region at the 5'-end of the mRNA and the sequence for the first 18 amino acids of preuteroglobin. The nucleotide sequence confirms most of the amino acid sequence of the signal peptide of uteroglobin as previously reported (Atger *et al.*, 1979; Malsky *et al.*, 1979), and helps clarify their contradictions. For example, at positions − 16 and − 11, there is threonine and at position − 6, there is cystein as found by Malsky *et al.* (1979).

5. The first exon is separated from the second one by a 2.3-kb–long intron. The second exon made up of 187 nucleotides has been completely sequenced and encodes the last three amino acids of the signal peptide of preuteroglobin and the first 60 amino acids of the mature uteroglobin. Between alanine 1 and glycine 1, there is a Pst1 site, thus explaining some of the difficulties we had in obtaining complete cDNA clones. As the cDNA was cloned in the Pst1 site of pBR322, excision with this enzyme leads to digestion of the cDNA into a large fragment containing the right two-thirds and a smaller fragment containing sequences of the left one-third.

6. The second exon is separated from the third exon by a 400-nucleotide-long intron, that has been sequenced. The third exon is comprised of 171 nucleotides and encodes the last 10 amino acids of uteroglobin and 140 nucleotides of the non-translated region at the 3'-end of the messenger RNA. The polyadenylation signal, AATAAA, is found 19 nucleotides before the end of the mRNA.

7. All the sequenced intron–exon transitions agree with the canonical sequence postulated by Breathnach *et al.* (1978) which has been found in most other known eukaryotic genes, namely, the limits of the introns were GT–AG.

8. An attempt to correlate the three exons of the gene with functional domains of the protein is difficult, since the structure of the steroid-binding site has not been elucidated, and other functions of uteroglobin are unknown. It is clear that the first exon contains the information for most of the signal peptide and that the main portion of the mature protein is encoded in the second exon. The last exon encodes a region of the protein that could represent a different secondary structure. In fact, in all secondary structure predictions, there is a strong domain of α-helical structure extending to amino acids 60–61, followed by the last nine or ten amino acids with much lower α-helical potential. According to X-ray crystallographic analysis of native oxidized uteroglobin, however, the α-helix extends to amino acid 65 (Mornon *et al.*, 1980). There is, therefore, no evident correlation between exons and structural or functional domains of uteroglobin as has been hypothesized (Gilbert, 1978) and

found for other proteins, including globins, immunoglobins, and lysozymes (Blake, 1979; Sakano *et al.*, 1979; Eaton, 1980; Jung *et al.*, 1980).

11. Concluding Remarks

In the few years since the identification of uteroglobin as a hormone-responsive protein, its induction has developed into an attractive system for the study of regulated gene expression. One limitation that this system shares with some other popular systems used for studying the mechanism of hormone action is the lack of a responsive cell culture from target cells. Such a culture system is needed not only to study in more detail the kinetics of induction, but also to provide a host cell for gene transfer experiments. Analysis of the expression and regulation of *in vitro*-manipulated uteroglobin genes after episomal transfer into a target cell would provide invaluable information for understanding the molecular mechanism of hormone action.

With the purified receptor and the isolated gene, binding experiments can be performed that may serve to identify sequences relevant for hormone regulation on or around the uteroglobin gene. Of course, this type of study may have to include reconstitution experiments with nucleosomes and other chromatin components, and would also profit from site-specific manipulations of the uteroglobin gene.

Another direction in which considerable effort is needed in the near future is the development of *in vitro* transcription systems that efficiently mimic all steps in the transcription and processing of RNA from structural genes. Such systems, whether based on isolated nuclei, purified chromatin, or nicked DNA, would allow an investigation of the role played by the hormone receptor during the production of specific mRNA. Certainly the clarification of this interaction will ultimately provide the answer to the question of the molecular mechanism by which hormones modulate gene expression.

ACKNOWLEDGMENTS. We thank Dr. A. C. Cato for his help and suggestions during the preparation of the manuscript, and the Deutsche Forschungsgemeinschaft for financial support (SFB 103-B1).

REFERENCES

Arnemann, J., Heins, B., and Beato, M., 1977, Purification and properties of rabbit uterus preuteroglobin mRNA, *Nucleic Acid Res.* **4**:4023–4036.
Arnemann, J., Heins, B., and Beato, M., 1979, Synthesis and characterization of a DNA complementary to preuteroglobin mRNA, *Eur. J. Biochem.* **99**:361–367.

Arthur, A. T. and Daniel, J. C., 1972, Progesterone regulation of blastokinin production and maintenance of rabbit blastocysts transferred into uteri of castrated recipients, *Fertil. Steril.* **23:**115–122.

Arthur, A. T., Cowan, B. D., and Daniel, J. C., 1972, Steroid binding to blastokinin, *Fertil. Steril.* **23:**85–92.

Atger, M., and Milgrom, E., 1977, Progesterone-induced mRNA. Translation, purification, and preliminary characterization of uteroglobin mRNA, *J. Biol. Chem.* **252:**5412–5418.

Atger, M., Mercier, J. C., Haze, G., Fridlansky, F., and Milgrom, E., 1979, N-Terminal sequence of uteroglobin and its precursor, *Biochem. J.* **177:**985–988.

Atger, M., Perricaudet, M., Tiollais, P., and Milgrom, E., 1980, Bacterial cloning of the rabbit uteroglobin structural gene, *Biochem. Biophys. Res. Commun.* **93:**1082–1088.

Atger, M., Atger, P., Tiollais, P., and Milgrom, E., 1981, Cloning of rabbit genomic fragments containing the uteroglobin gene, *J. Biol. Chem.* **256:**5970–5972.

Barfield, M. A., Stambaugh, R., Mastroianni, L., and Storey, B. T., 1976, Factors in diminution of uteroglobin secretion in the rabbit, *Fertil. Steril.* **27:**39–45.

Beato, M., 1976, Binding of steroids of uteroglobin, *J. Steroid Biochem.* **7:**327–334.

Beato, M., 1977a, Hormonal control of uteroglobin biosynthesis, in: *Development in Mammals,* Vol. I (M. H. Johnson, ed.), pp. 361–384, North Holland Publishing Co. Amsterdam.

Beato, M., 1977b, Physico-chemical characterization of uteroglobin and its interaction with progesterone, in: *Development in Mammals,* Vol. II (M. H. Johnson, ed.), pp. 173–198, North Holland Publishing Co. Amsterdam.

Beato, M., and Arnemann, J., 1975, Hormone-dependent synthesis and secretion of uteroglobin in isolated rabbit uterus, *FEBS Lett.* **58:**126–129.

Beato, M., and Baier, R., 1975, Binding of progesterone to the proteins of the uterine luminal fluid. Identification of uteroglobin as the binding protein, *Biochim. Biophys. Acta* **392:** 346–356.

Beato, M., and Beier, H. M., 1978, Characteristics of the purified uteroglobin-like protein from rabbit lung, *J. Reprod. Fertil.* **53:**305–314.

Beato, M., and Nieto, A., 1976, Translation of the mRNA for rabbit uteroglobin in cell-free systems. Evidence for a precursor protein, *Eur. J. Biochem.* **64:**15–21.

Beato, M. and Runger, D., 1975, Translation of the messenger RNA for rabbit uteroglobin in *Xenopus* oocytes, *FEBS Lett.* **59:**305–309.

Beato, M., Arnemann, J., and Voss, H-J., 1977, Spetrophotometric study of progesterone binding to uteroglobin, *J. Steroid Biochem.* **8:**725–730.

Beato, M., Saavedra, A., Puigdomenech, P., Tancredi, T., and Temussi, P. A., 1980a, Progesterone binding to uteroglobin, in: *Steroid Induced Uterine Proteins* (M. Beato, ed.), pp. 105–119, Elsevier North-Holland Biomedical Press, Amsterdam.

Beato, M., Arnemann, J., Heins, B., Müller, H., and Nieto, A., 1980b, Correlation between uteroglobin synthesis and uteroglobin mRNA content in rabbit endometrium, in: *Steroid Induced Uterine Proteins* (M. Beato, ed.), pp. 351–368, Elsevier North-Holland Biomedical Press, Amsterdam.

Beier, H. M., 1968, Uteroglobin: A hormone-sensitive endometrial protein involved in blastocyst development, *Biochim. Biophys. Acta* **160:**289–291.

Beier, H. M., Kühnel, W., and Petry, G., 1971, Uterine secretion proteins as extrinsic factors in preimplantation development, *Adv. Biosci.* **6:**165–189.

Beier, H. M., Bohn, H., and Müller, W., 1975, Uteroglobin-like antigen in the male genital tract secretions, *Cell Tissue Res.* **165:**1–11.

Beier, H. M., Mootz, U., and Fischer, B., 1980, New aspects on the physiology of uteroglobin, in: *Steroid Induced Uterine Proteins* (M. Beato, ed.), pp. 47–67, Elsevier/North-Holland Biomedical Press, Amsterdam.

Benton, W. D., and Davis, R. W., 1977, Screening λgt recombinant clones by hybridization to single plaques *in situ, Science* **196:**180–182.

Berliner, J. A., and Gerschenson, L. E., 1976, Sex steroid induced morphological changes in primary uterine cell cultures, *J. Steroid Biochem.* **7:**153–158.

Bernstein, G. S., Aladjem, F., and Chen, S., 1971, Proteins in human endometrial washings—a preliminary report, *Fertil. Steril.* **22:**722–726.

Blake, C. C. F., 1979, Exons encode protein functional units, *Nature* **277:**598.

Bochskanl, R., and Kirchner, C., 1981, Uteroglobin and the accumulation of progesterone in the uterine lumen of the rabbit, *Arch. Dev. Biol.* **190:**127–131.

Breathnach, R., Benoist, C., O'Hare, K., Gannon, F., and Chambon, P., 1978, Ovalbumin gene: Evidence for a leader sequence in mRNA and DNA sequences at the exon–intron boundaries, *Proc. Natl. Acad. Sci. USA* **75:**4853–4857.

Buehner, M., and Beato, M., 1978, Crystallization and preliminary crystallographic data of rabbit uteroglobin, *J. Mol. Biol.* **120:**337–341.

Bullock, D. W., 1977a, Progesterone induction of mRNA and protein synthesis in the rabbit uterus, *Ann. N.Y. Acad. Sci.* **286:**260–272.

Bullock, D. W., 1977b, *In vitro* translation of mRNA for a uteroglobin-like protein from rabbit lung, *Biol. Reprod.* **17:**104–107.

Bullock, D. W., and Connell, K. M., 1973, Occurrence and molecular weight of rabbit uterine "blastokinin," *Biol. Reprod.* **9:**125–132.

Bullock, D. W., and Willen, G. F., 1974, Regulation of a specific uterine protein by estrogen and progesterone in ovariectomized rabbits, *Proc. Soc. Exp. Biol. Med.* **146:**294–298.

Bullock, D. W., Woo, S. L. C., and O'Malley, B. W., 1976, Uteroglobin mRNA: Translation *in vitro, Biol. Reprod.* **15:**435–443.

Chandra, T., Woo, S. L. C., and Bullock, D. W., 1980, Cloning of the rabbit uteroglobin structural gene, *Biochem. Biophys. Res. Commun.* **95:**197–204.

Chen, L., Lindner, H. R., and Lancet, M. Y., 1973, Mitogenic action of oestradiol-17β on human myometrial and endometrial cells in long-term tissue culture, *J. Endocrinol.* **59:**87–93.

Daniel, J. C. Jr., 1964, Early growth of rabbit trophoblast, *Am. Natural.* **98:**85–97.

Daniel, J. C. Jr., 1971, Uterine proteins and embryonic development, *Adv. Biosci.* **6:**191–203.

Daniel, J. C. Jr., 1972, A blastokinin-like component from the human uterus, *Fertil. Steril.* **23:**522–528.

Daniel, J. C., and Levy, J. D., 1964, Action of progesterone as a cleavage inhibitor of rabbit ova *in vitro, J. Reprod. Fertil.* **7:**323–329.

Daniel, J. C. Jr., and Milazzo, J. T., 1976, Continuity of a rabbit antigen between generations, *Cancer Res.* **36:**3409–3411.

Eaton, W. A., 1980, The relationship between coding sequences and function in haemoglobin, *Nature* **284:**183–185.

El-Banna, A. A., and Daniel, J. C., 1972, Stimulation of rabbit blastocysts *in vitro* by progesterone and uterine proteins in combination, *Fertil. Steril.* **23:**101–104.

El-Banna, A. A., and Daniel, J. C., 1972b, The effect of protein fractions from rabbit uterine fluids on embryo growth and uptake of nucleic acid and protein precursors, *Fertil. Steril.* **23:**105–114.

El-Banna, A. A., and Sacher, B., 1977, A study on steroid hormone receptors in the rabbit oviduct and uterus during the first few days after coitus and during egg transport, *Biol. Reprod.* **17:**1–8.

Faber, L. E., Sandmann, M. L., and Stavely, H. E., 1972, Progesterone-binding proteins of the rat and rabbit uterus, *J. Biol. Chem.* **247:**5648–5649.

Faber, L. E., Sandmann, M. L., and Stavely, H. E., 1973, Progesterone and corticosterone binding in rabbit uterine cytosols, *Endocrinology* **93:**74–81.

Feigelson, M., Noske, I. G., Goswami, A. K., and Kay, E., 1977, Reproduction tract fluid proteins and their hormonal control, *Ann. N.Y. Acad. Sci.* **286:**273–286.

Fleischmann, G., and Beato, M., 1978, Characterization of the progesterone receptor of rabbit uterus with the synthetic progestin 16α-ethyl-21-hydroxy-19-norpregn-4-ene-3,20-dione, *Biochim. Biophys. Acta* **540:**500–517.

Fleischmann, G., and Beato, M., 1979, Activation of the progesterone receptor of rabbit uterus, *Mol. Cell. Endocrinol.* **16:**181–197.

Fridlansky, F., and Milgrom, E., 1976, Interaction of uteroglobin with progesterone, 5α-pregnane-3,20-dione and estrogens, *Endocrinology* **99**:1244–1251.

Gerschenson, L. E., and Berliner, J. A., 1976, Further studies on the regulation of cultured rabbit endometrial cells by diethylstilbestrol and progesterone, *J. Steroid Biochem.* **7**:159–165.

Gerschenson, L. E., Berliner, J., and Yang, J. J., 1974, Diethylstilbestrol and progesterone regulation of cultured rabbit endometrial cell growth, *Cancer Res.* **34**:2873–2880.

Gerschenson, L. E., Conner, E., and Murai, J. T., 1977, Regulation of the cell cycle by diethylstilbestrol and progesterone in cultured endometrial cells, *Endocrinology* **100**:1468–1471.

Gilbert, W., 1978, Why genes in pieces? *Nature* **271**:501.

Goswami, A., and Feigelson, M., 1974, Differential regulation of a low-molecular weight protein in oviductal and uterine fluids by ovarian hormones, *Endocrinology* **95**:669–675.

Grunstein, M., and Hogness, D. S., 1975, Colony hybridization: A method for the isolation of cloned DNAs that contain a specific gene, *Proc. Natl. Acad. Sci. USA* **72**:3961–3965.

Gulyas, B. J., Daniel, J. C., and Krishnan, R. S., 1969, Incorporation of labeled nucleosides *in vitro* by rabbit and mink blastocyst in the presence of blastokinin or maternal rabbit serum, *J. Reprod. Fertil.* **20**:256–262.

Heins, B., and Beato, M., 1981, Hormonal control of uteroglobin secretion and preuteroglobin mRNA content in rabbit endometrium, *Mol. Cell. Endocrinol.* **21**:139–150.

Hemminki, S. M., Kopu, H. T., Torkkeli, T. K., and Jänne, O. A., 1980, Further studies on the role of estradiol in the induction of progesterone-regulated uteroglobin synthesis in the rabbit uterus, *Mol. Cell. Endocrinol.* **17**:71–80.

Hilliard, J., and Eaton, L. W., 1971, Estradiol-17β, progesterone and 20α-hydroxyprogesterone in rabbit ovarian venous plasma. II. From mating through implantation, *Endocrinology* **89**:522–527.

Hossner, K. L., 1976, An investigation of blastokinin–progesterone interaction *in vitro,* Ph.D. dissertation, University of Tennessee, Knoxville.

Isomaa, V., Isotalo, H., Orava, M., and Jänne, O., 1979a, Regulation of cytosol and nuclear progesterone receptors in rabbit uterus by estrogen, antiestrogen and progesterone administration, *Biochim. Biophys. Acta* **585**:24–33.

Isomaa, V., Isotalo, H., Orava, M., Torkkeli, T., and Jänne, O., 1979b, Changes in cytosol and nuclear progesterone receptor concentrations in the rabbit uterus and their relation to induction of progesterone-regulated uteroglobin, *Biochem. Biophys. Res. Commun.* **88**:1237–1243.

Isotalo, H., Isomaa, V., and Jänne, O., 1981, Replenishment and properties of cytosol progesterone receptor in rabbit uterus after multiple progesterone administrations, *Endocrinology* **108**:868–873.

Johnson, M. H., 1972, The distribution of a blastokinin-like uterine protein studied by immune fluorescence, *Fertil. Steril.* **23**:929–939.

Joshi, S. G., and Ebert, K. M., 1976, Effect of progesterone on labelling of soluble proteins and glycoproteins in rabbit endometrium, *Fertil. Steril.* **27**:730–739.

Jung, A., Sippel, A. E., Grez, M., and Schütz, G., 1980, Exons encode functional and structural units of chicken lysozyme, *Proc. Natl. Acad. Sci. USA* **77**:5759–5763.

Karlson, P., 1961, Biochemische Wirkungsweise der Hormone, *Dtsch. Med. Wochenschr.* **86**:668–674.

Kay, E., and Feigelson, M., 1972, An estrogen modulated protein in rabbit oviductal fluid, *Biochim. Biophys. Acta* **271**:436–441.

Kirchner, C., 1972, Immune histologic studies on the synthesis of a uterine specific protein in the rabbit and its passage through the blastocyst coverings, *Fertil. Steril.* **23**:131–136.

Kirchner, C., 1976a, Uteroglobin in the rabbit. I. Intracellular localization in the oviduct, uterus and preimplantation blastocyst, *Cell Tissue Res.* **170**:415–424.

Kirchner, C., 1976b, Uteroglobin in the rabbit. II. Intracellular localization in the uterus after hormone treatment *Cell Tissue Res.* **170**:425–434.

Kirchner, C., 1979, Immunohistochemical localization of secretory proteins in the endometrial epithelium of the rabbit, 1979, *Cell Tissue Res.* **199**:25–36.

Kirchner, C., 1980, Non-uteroglobin proteins in the rabbit, in: *Steroid Induced Uterine Proteins* (M. Beato, ed.), pp. 69–86, Elsevier/North-Holland Biomedical Press, Amsterdam.

Kirchner, C., and Schroer, H. G., 1976, Uterine secretion-like proteins in the seminal plasma of the rabbit, *J. Reprod. Fertil.* **47**:325–330.

Kokko, E., Isomaa, V., and Jänne, O., 1977, Progesterone-regulated changes in transcriptional events in rabbit uterus, *Biochim. Biophys. Acta* **479**:354–366.

Kopu, H. T., Hemminki, S. M., Torkkeli, T. K., and Jänne, O., 1979, Hormonal control of uteroglobin secretion in rabbit uterus. Inhibition of uteroglobin synthesis and mRNA accumulation by estrogen and anti-estrogen administration, *Biochem. J.* **180**:491–500.

Krishnan, R. S., 1970, Incorporation *in vivo* of ^{14}C-labeled precursors into the protein of rabbit uterine secretions, *Arch. Biochem. Biophys.* **141**:764–765.

Krishnan, R. S., and Daniel, J. C. Jr., 1967, "Blastokinin": Inducer and regulator of blastocyst development in the rabbit uterus, *Science* **158**:490–492.

Levey, I. L., and Daniel, J. C. Jr., 1976, Isolation and translation of blastokinin mRNA, *Biol. Reprod.* **14**:163–174.

Logeat, F., Hai, M. T. V., and Milgrom, E., 1981, Antibodies to rabbit progesterone receptor: Crossreaction with human receptor, *Proc. Natl. Acad. Sci. USA* **78**:1426–1430.

Loosfelt, H., Fridlansky, F., Savouret, J. F., Atger, M., and Milgrom, E., 1981, Mechanism of action of progesterone in the rabbit endometrium. Induction of uteroglobin and its mRNA, *J. Biol. Chem.* **256**:3465–3470.

Malsky, M. L., Bullock, D. W., Willard, J. J., and Ward, D. N., 1979, Progesterone-induced secretory protein. NH$_2$-terminal sequence of pre-uteroglobin, *J. Biol. Chem.* **254**:1580–1585.

Maniatis, T., Hardison, R. C., Lacy, E., Lauer, J., O'Connel, C., Quon, D., Sim, G. K., and Efstratiadis, A., 1978, The isolation of structural genes from libraries of eucaryotic DNA, *Cell* **15**:687–701.

Maurer, R. R., and Beier, H. M., 1976, Maternal aging and embryonic mortality in the rabbit. II. Hormonal changes in young and aging females, *J. Reprod. Fertil.* **31**:15–22.

Maxam, A., and Gilbert, W., 1980, Sequencing end-labeled DNA with base specific chemical cleavages, *Methods Enzymol.* **65**:499–560.

Mayol, R. F., and Longenecker, D. E., 1974, Development of a radioimmunoassay for blastokinin, *Endocrinology* **95**:1534–1542.

McGaughey, R. W., and Murray, F. A., 1972, Properties of blastokinin: Amino acid composition, evidence for subunits, and estimation of isoelectric point, *Fertil. Steril.* **23**:399–404.

McGuire, J. L., and Bariso, C. D., 1972, Isolation and preliminary characterization of a progesterone specific binding macromolecule from the 273,000 g supernatant of rat and rabbit uteri, *Endocrinology* **90**:496–501.

McGuire, J. L., Bariso, C. D., and Shroff, A. P., 1974, Interaction between steroids and a uterine progestagen specific binding macromolecule, *Biochemistry* **13**:319–322.

Menne, C., Suske, G., Arnemann, J., Wenz, M., Cato, A. C. B., and Beato, M., 1982, Isolation and structure of the gene for the progesterone-inducible protein uteroglobin, *Proc. Natl. Acad. Sci. USA* **79**:4853–4857.

Mirault, M. E., Goldschmidt-Clermont, M., Artavanis-Bakonas, S., and Schedl, P., 1979, Organization of the multiple genes for the 70,000-dalton hat-shock protein in *Drosophila melanogaster*, *Proc. Natl. Acad. Sci. USA* **76**:5254–5258.

Mornon, J. P., Surcouf, E., Bally, R., Fridlansky, F., and Milgrom, E., 1978, X-ray analysis of a progesterone-binding protein (uteroglobin): Preliminary results, *J. Mol. Biol.* **122**:237–239.

Mornon, J. P., Fridlansky, F., Bally, R., and Milgrom, E., 1979, Characterization of two new crystal forms of uteroglobin, *J. Mol. Biol.* **127**:237–240.

Mornon, J. P., Fridlansky, F., Bally, R., and Milgrom, E., 1980, X-ray crystallographic analysis of a progesterone-binding protein. The C222$_1$ crystal form of oxidized uteroglobin at 2.2 Å resolution, *J. Mol. Biol.* **137**:415–429.

Mukherjee, A. B., Laki, K., and Agrawal, A. K., 1980, Possible mechanism of success of an allotransplantation in nature: Mammalian pregnancy, *Med. Hypotheses* **6**:1043–1055.

174

Miguel Beato et al.

Müller, H., and Beato, M., 1980, RNA synthesis in rabbit endometrial nuclei. Hormonal regulation of transcription of the uteroglobin gene, *Eur. J. Biochem.* **112**:235–241.

Murai, J. T., Conner, E., Yang, J., Andersson, M., Berliner, J., and Gerschenson, L. E., 1978, Hormonal regulation of cultured rabbit endometrial cells, *J. Toxicol. Environ. Health* **4**:449–456.

Murray, F. A., and Daniel, J. C. Jr., 1973, Synthetic pattern of proteins in rabbit uterine flushings, *Fertil. Steril.* **24**:692–697.

Murray, F. A., McGaughey, R. W., and Yarus, M. J., 1972, Blastokinin: Its size and shape, and an indication of the existence of subunits, *Fertil. Steril.* **23**:69–77.

Neulen, J., Beato, M., and Beier, H. M., 1982, Cytosol and nuclear progesterone receptor concentrations in the rabbit endometrium during early pseudopregnancy under different treatments with estradiol and progesterone, *Mol. Cell. Endocrinol.* **25**:183–191.

Nieto, A., and Beato, M., 1980, Synthesis and secretion of uteroglobin in rabbit endometrial explants cultured *in vitro, Mol. Cell. Endocrinol.* **17**:25–39.

Nieto, A., Ponstingl, H., and Beato, M., 1977, Purification and quaternary structure of the hormonally induced protein uteroglobin, *Arch. Biochem. Biophys.* **180**:82–92.

Norgard, M. V., Keem, K., and Monahan, J., 1978, Factor affecting the transformation of *E. coli* strain χ1776 by pBR322 plasmid DNA, *Gene* **3**:279–292.

Noske, I. G., and Feigelson, M., 1976, Immunological evidence of uteroglobin (blastokinin) in the male reproductive tract and in nonreproductive ductal tissues and their secretions, *Biol. Reprod.* **15**:704–713.

Orava, M. M., Isomaa, V. V., and Jänne, O. A., 1979, Nuclear poly(A) polymerase activities in the rabbit uterus, *Eur. J. Biochem.* **101**:195–203.

Pavlik, E. J., and Katzenellenbogen, B. S., 1978, Human endometrial cells in primary tissue culture: Estrogen interactions and modulation of cell proliferation, *J. Clin. Endocrinol.* **47**:333–334.

Petzoldt, U., Dames, W., Gottschweski, G. H. M., and Neuhoff, V., 1972, Das Proteinmuster in frühen Entwicklungsstadien des Kaninchens, *Cytobiology* **5**:272–280.

Philibert, D., and Raynaud, J. P., 1974, Progesterone binding in the immature rabbit and guinea pig uterus, *Endocrinology* **94**:627–632.

Philibert, D., Ojasoo, T., and Raynaud, J. P., 1977, Properties of the cytoplasmic progestin-binding protein in the rabbit uterus, *Endocrinology* **101**:1850–1861.

Ponstingl, H., Nieto, A., and Beato, M., 1978, Amino acid sequence of progesterone-induced rabbit uteroglobin, *Biochemistry* **17**:3908–3912.

Popp, R. A., Foresman, K. R., Wise, L. D., and Daniel, J. C. Jr., 1978, Amino acid sequence of a progesterone-binding protein, *Proc. Natl. Acad. Sci. USA* **75**:5516–5519.

Puigdoménech, P., and Beato, M., 1977, Nuclear magnetic resonance studies on rabbit uteroglobin, *FEBS Letts.* **83**:217–221.

Rahman, S. S. U., Billiar, R. B., and Little, B., 1975a, Induction of uteroglobin in rabbits by progestogens, estradiol-17β and ACTH, *Biol. Reprod.* **12**:305–314.

Rahman, S. S. U., Velayo, N., Domres, P., and Billiar, R. B., 1975b, Evaluation of progesterone binding to uteroglobin, *Fertil. Steril.* **26**:991–995.

Rahman, S. S., Billiard, R. B., and Little, B., 1981, Studies of the decline of uteroglobin synthesis and secretion in the rabbit uterus during the continued presence of circulatius progesterone, *Endocrinology* **18**:2222–2227.

Rao, B. R., and Katz, R. M., 1977, Progesterone receptors in rabbit uterus. II. Characterization and estrogen augmentation, *J. Steroid Biochem.* **8**:1213–1220.

Rao, B. R., Wiest, W. G., and Allen, W. M., 1973, Progesterone receptor in rabbit uterus. I. Characterization and estradiol-17β augmentation, *Endocrinology* **92**:1229–1240.

Rigby, P. W. J., Dieckmann, M., Rhodes, C., and Berg, P., 1977, Labeling DNA to high specific activity *in vitro* by nick translation with DNA polymerase I, *J. Mol. Biol.* **113**:237–251.

Roberts, B. E., and Paterson, B. M., 1973, Efficient translation of TMV RNA and rabbit globin as RNA in a cell-free system from commercial wheat germ, *Proc. Natl. Acad. Sci. USA* **70**:2330–2334.

Roblero, L., 1973, Effect of progesterone *in vivo* upon rate of cleavage of mouse embryos, *J. Reprod. Fertil.* **35**:153–158.

Saavedra, A., and Beato, M., 1980, Influence of chemical modifications of amino acid side chains on the binding of progesterone to uteroglobin, *J. Steroid Biochem.* **13**:1347–1353.

Sakano, H., Rogers, J. H., Hüppi, K., Brack, Ch., Traunecker, A., Maki, R., Wall, R., and Tonegawa, S., 1979, Domains and the hinge region of an immunoglobulin heavy chain are encoded in separate DNA segments, *Nature* **277**:627–633.

Savouret, J. F., Loosfelt, H., Atger, M., and Milgrom, E., 1980, Differential hormonal control of a mRNA in two tissues. Uteroglobin mRNA in the lung and the endometrium, *J. Biol. Chem.* **255**:4131–4136.

Schütz, G., Beato, M., and Feigelson, P., 1974, Isolation on cellulose of ovalbumin and globin mRNA and their translation in an ascites cell-free system, *Methods Enzymol.* **30**:701–708.

Shirai, E., Iizuka, R., and Notake, Y., 1972, Analysis of human uterine fluid proteins, *Fertil. Steril.* **23**:522–528.

Sonnenschein, C., Weiller, S., Farookhi, R., and Soto, A. M., 1974, Characterization of an estrogen-sensitive cell line established from normal rat endometrium, *Cancer Res.* **34**:3147–3153.

Tamaya, T., Nioka, S., Furuta, N., Shimura, T., Takano, N., and Okada, H., 1977, Contribution of functional groups of 19-Nor-progestogens to binding to progesterone and estradiol-17β receptor in rabbit uterus, *Endocrinology* **100**:1579–1584.

Tancredi, T., Temussi, P. A., and Beato, M., 1982, Interaction of oxydized and reduced uteroglobin with progesterone, *Eur. J. Biochem.* **122**:101–104.

Temussi, P. A., Tancredi, T., Puigdoménech, P., Saavedra, A., and Beato, M., 1980, Interaction of S-carboxymethylated uteroglobin with progesterone, *Biochemistry* **19**:3287–3293.

Torkkeli, T., 1980, Early changes in rabbit uterine progesterone receptor concentrations and uteroglobin synthesis after progesterone administration, *Biochem. Biophys. Res. Commun.* **97**:559–565.

Torkkeli, T. K., Krusius, T., and Jänne, O. A., 1978, Uterine and lung uteroglobin in the rabbit: Two identical proteins with differential hormonal regulation, *Biochim. Biophys. Acta* **544**:578–592.

Torkkeli, T. K., Kontula, K. K., and Jänne, O. A., 1977, Hormonal regulation of uterine blastokinin synthesis and occurrence of blastokinin-like antigens in non-uterine tissues. *Mol. Cell. Endocrinol.* **9**:101–118.

Urzua, M. A., Stambaugh, R., Flickinger, G., and Mastroianni, L. Jr., 1970, Uterine and oviduct fluid protein patterns in the rabbit before and after ovulation, *Fertil. Steril.* **21**:860–865.

Voss, H-J., and Beato, M., 1977, Human fluid proteins: Gel electrophoretic pattern and progesterone-binding proteins, *Fertil. Steril.* **28**:972–980.

Westphal, H. W., Fleischmann, G., Climent, F., and Beato, M., 1978, Effect of phospholipases and lysophosphatides on partially purified steroid hormone receptors, Hoppe-Seyler's *Z. Physiol. Chem.* **359**:1297–1305.

Westphal, H. M., Fleischmann, G., and Beato, M., 1981, Photoaffinity labeling of steroid binding proteins with unmodified ligands, *Eur. J. Biochem.* **119**:101–106.

Whitson, G. L., and Murray, F. A., 1974, Cell culture of mammalian endometrium and synthesis of blastokinin *in vitro, Science* **183**:668–669.

Whitten, W. K., 1957, The effect of progesterone on the development of mouse eggs *in vitro, J. Endocrinol.* **16**:80–84.

Wolf, D. P., and Mastroianni, L. Jr., 1975, Protein composition of human uterine fluid, *Fertil. Steril.,* **26**:240–245.

Young, C. E., Smith, R. G., and Bullock, D. W., 1980, Uteroglobin mRNA activity and levels of nuclear progesterone receptor in endometrium, *Mol. Cell. Endocrinol.* **20**:219–226.

9

Evolution and Regulation of Genes for Growth Hormone and Prolactin

Walter L. Miller and Synthia H. Mellon

1. Evolution

1.1. Introduction

Prolactin (Prl), growth hormone (GH), and chorionic somatomammotropin (CS, placental lactogen) form a set of related polypeptide hormones that appear to have derived from a common evolutionary ancestor protein (Catt *et al.*, 1967; Sherwood, 1967; Li *et al.*, 1967; Niall *et al.*, 1971). They are related by function, immunochemistry, and structure. All are lactogenic and growth-promoting, they are all of similar size (190–199 amino acids among various species), and they all have similar protein structures. Each hormone has a single homologous tryptophan residue at about locus 85 (GH and CS) or 91 (Prl), and two homologous disulfide bonds. The three hormones also each contain four internal regions of homology which are themselves homologous among the three hormones. Based on these observations from the amino acid sequencing data available in 1971, Niall *et al.* (1971) postulated that the three hormones had arisen by duplication of an ancestral hormone gene. Recombinant DNA technology now permits the detailed analysis of the gene sequences encoding these hormones, permitting reevaluation of this hypothesis from data describing the gene sequences themselves.

Walter L. Miller ● Department of Pediatrics and Metabolic Research Unit, Department of Medicine, University of California, San Francisco, California 94143   *Synthia H. Mellon* ● Metabolic Research Unit, Department of Medicine, University of California, San Francisco, California 94143

To examine the evolution of this related set of genes we can make two types of comparisons: we can compare the genes for a single hormone in several species and we can compare the genes for the three related hormones in a single species. At the present time complete cDNA sequence data exist for bovine (Miller *et al.*, 1980), human (Martial *et al.*, 1979), and rat (Seeburg *et al.*, 1977) growth hormone; bovine (Miller *et al.*, 1981, 1982; Sasavage *et al.*, 1982), human (Cooke *et al.*, 1981), and rat (Gubbins *et al.*, 1980; Cooke *et al.*, 1980) prolactin; and human chorionic somatomammotropin (Shine *et al.*, 1977; Goodman *et al.*, 1980). Comparison of cDNA sequences is useful for examining rates of nucleotide changes that cause one amino acid to be replaced by another (replacement mutations), and nucleotide changes that cause no amino acid change (silent changes). Examination of cDNA sequences, however, only reveals changes in those portions of a gene that eventually are processed to a mature mRNA molecule. This is more informative than amino acid sequence data, as it includes the information in the leader peptide and untranslated regions of the mRNAs. A second useful procedure entails examining the structure and organization of the genes themselves. Cloned genomic DNA reveals information about gene size, arrangement, and origins, providing a macroscopic counterpoint revealing earlier evolutionary events in contrast with the detailed analysis generally done with cDNA.

1.2. Structure of Cloned cDNAs

Before examining the structures and evolution of the protein hormones themselves, we shall first illustrate the kinds of information available from cloned cDNAs which cannot be determined by amino acid analysis. A general review of recombinant DNA technology has been published recently (Miller, 1981). All but two of the amino acids may be encoded by more than one codon (triplet of nucleotides). Where such "redundancy" exists in the genetic code, the identity of the third nucleotide in the codon generally may vary. Since cDNA is an enzymatic copy of the mRNA, analysis of cDNA sequences will reveal which codons are used in a particular gene. A simple way of quantifying codon choice is simply to count the number of codons ending in G or C as opposed to those ending in A or T. Because of the specifics of the genetic code and the fact that certain amino acids are used rarely, random codon selection would result in 42% of codons ending in G or C. However, all of the growth hormone and prolactin genes have a preference for codons ending in G or C ranging from 50 to 82% (Table I). Codon choice does not appear to be a species-related phenomenon, as shown in Table I. All three GHs and hCS exhibit a strong preference for G and C (74–82%), while the three Prls have a lesser preference (50–63%). Codon choice may be determined by the need to establish or prohibit the formation of secondary structures, but this is unproven (Miller *et al.*, 1981).

The sequence of a cDNA may also reveal the structure of the untranslated

Table I. Relative Abundances (%) of Codons
Ending in G or C in Various Hormone mRNAs[a]

Prl/GH-related hormones		Bovine hormones	
bGH	82	Growth hormone	82
hCS	80	Proopiomelanocortin	65
hGH	76	Prolactin	60
rGH	74	Parathyroid hormone	43
hPrl	63		
bPrl	60		
rPrl	50		

[a] Data are derived from published sequences: bGH, Miller *et al.*
(1980); hCS, Shine *et al.* (1977), and Goodman *et al.* (1980);
hGH, Martial *et al.* (1979); rGH, Seeburg *et al.* (1977); hPrl,
Cooke *et al.* (1981); rPrl, Cooke *et al.* (1980); bovine
proopiomelanocortin (the precursor to adrenocorticotropin and β-
endorphin), Nakanishi *et al.* (1979); bovine (pre-pro-) parathyroid
hormone, Kronenberg *et al.* (1979).

portions of the mRNA. These untranslated regions occur at the 5' (proximal)
end of the mRNA, where they are involved in binding the mRNA to ribosomes
and in regulating translation, and they occur at the 3' (distal) end, where their
function is unknown. Upon examining the available sequences of the 5' untranslated
regions of the seven cloned hormones (Martial *et al.*, 1979; Seeburg *et al.*, 1977;
Miller *et al.*, 1980; Sasavage *et al.*, 1982; Cooke *et al.*, 1981; Gubbins *et al.*,
1980; Shine *et al.*, 1977; Goodman *et al.*, 1980), obvious homologies are not
apparent except between hGH and hCS, where 26 of the 29 known nucleotides
are identical. Examination of the 3' untranslated regions is more rewarding. The
three prolactin mRNAs appear to form a group different from the GH and CS
mRNAs based on the structure of their 3' untranslated regions as well as on
their codon choice. The prolactin group uses UAA to terminate transcription
while the growth hormone group uses UAG. Following this termination codon,
the prolactin group is rich in A and U in the proximal region of the 3' untranslated
region (an average of 59% of the first 43 nucleotides), while the growth hormone
group is rich in G and C (69% of the first 47 nucleotides). The prolactin mRNAs
from all three species contain considerable homology only at the proximal end
(10 of 15 nucleotides) and in the region of the AAUAAA sequence which is
common to all known eukaryotic mRNAs (Proudfoot and Brownlee, 1976).
Among the growth hormone mRNAs from the three species there is little homology
in the proximal region but considerable homology elsewhere in addition to the
AAUAAA region. Finally, the 3' untranslated regions for the growth hormone
group contain long palindromic dyads rich in cytosine (Martial *et al.*, 1979;
Seeburg *et al.*, 1977; Shine *et al.*, 1977; Miller *et al.*, 1980), whereas only
bovine and human prolactin mRNAs have short U-rich palindromes respectively

centered 19 and 17 nucleotides from the UAA termination codons. Such palindromic dyads strongly favor a single-stranded structure for the 3' untranslated region, possibly implying a regulatory role in mRNA translation or processing. It may be noteworthy that although the prolactin and growth hormone genes are expressed in the same tissue, their expression tends to be reciprocal, i.e., growth hormone production frequently decreases when prolactin increases and vice versa. Hence on the basis of codon use and the structures of the untranslated regions of the mRNAs it appears that the prolactin and growth hormone constitute different groups, and that hCS appears to be a variant of hGH.

1.3. The Evolutionary Clock Hypothesis

Zuckerkandl and Pauling (1962, 1965) first noted that proteins generally appear to evolve at a constant rate. This was taken to indicate that mutations occur at a roughly constant rate, permitting the use of amino acid or nucleotide sequence homology data as an evolutionary clock. The classic way of comparing apparently related proteins is by aligning the amino acids and counting the number of amino acid differences. If mutations occur at an approximately constant rate in evolution, then the number of amino acids that are different in two related proteins will serve as an index of the evolutionary time elapsed since the genes for the two proteins diverged. Several procedures have been devised for comparing such amino acid homologies (Dayhoff *et al.*, 1972; Wilson *et al.*, 1977). The most widely used system is the unit evolutionary period (UEP), which directly incorporates the concept of the evolutionary clock (Wilson *et al.*, 1977). The UEP is simply the length of time (in millions of years) required for 1% of the amino acids to change in two related proteins. Some proteins appear to evolve more slowly than others. Comparison of amino acid homology data to paleontologically determined times of species divergence indicates that a 1% amino acid difference accumulates in cytochrome *c* every 15 million years (hence its UEP is 15), whereas 400 million years are required for a 1% change in histone H4. Hence the evolutionary clock runs much faster for cytochrome c than for histones. Wilson estimates that the UEPs for prolactin and growth hormone are 5.0 and 4.0, respectively (Wilson *et al.*, 1977). As we concur with the hypothesis that growth hormone and prolactin arose from a common evolutionary ancestor (Catt *et al.*, 1967; Sherwood, 1967; Li *et al.*, 1967; Niall *et al.*, 1971) and as comparisons between the two hormones are facilitated by identical UEPs, we use the value 4.5. This is also the value calculated for growth hormone and prolactin by Cooke *et al.* (1981) using the procedure of Perler *et al.* (1980).

If the evolutionary date of divergence of two peptides can be calculated by comparing the amino acid sequences in various species, i.e., if it is possible to generate an evolutionary clock, then it must be possible to calculate the time of evolutionary divergence of various species based on known amino acid sequences of common peptides of known UEP. In other words, the clock must apply to both the peptides and the species. Therefore, a valid evolutionary clock should be able to determine either the divergence pattern of the species (by comparing one gene or protein in several animals) or the divergence patterns of the genes (by comparing related genes or proteins in a single species).

1.4. Comparison of Sequences

To compare the nucleotide or amino acid sequences of prolatin and growth hormone among several species, gaps must be introduced in some sequences to permit alignment of homologous regions. Alignment of bovine, rat, and human prolactin or growth hormone was done by inspection (Fig. 1). However for comparison of prolactin to growth hormone, inspection identifies certain areas of homology, but does not reveal where the required gaps in the various sequences should lie. Others have introduced gaps with computer programs designed to maximize nucleotide sequence homologies (Perler *et al.,* 1980; Efstratiadis *et al.,* 1980). However, if evolutionary selective pressure acts primarily to choose favorable protein hormones rather than nucleotide sequences, evolutionary drift in nonexpressed sites might induce a computer program to assign gaps too arbitrarily without considering the impact of amino acid changes on protein configuration. Therefore, we introduced gaps in sequences by estimating the "acceptability" of possible amino acid pairings. To do this, we compared the relative frequency with which any given amino acid will replace an homologous amino acid in two related peptides. These relative frequencies have been determined empirically by examination of 250 amino acid replacements in various proteins (Dayhoff *et al.,* 1978). These relative frequencies correlate closely with amino

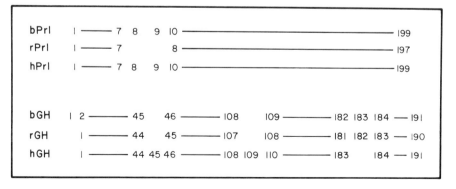

Figure 1. Top: Alignment of the amino acid and nucleotide sequences of bovine, rat, and human prolactins. Numbers refer to amino acid residues. Bottom: Alignment of bovine, rat, and human growth hormone sequences.

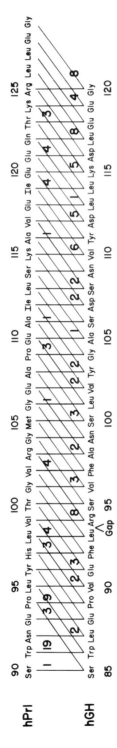

Figure 2. Procedure used to introduce gaps in homologous sequences. A portion of the human prolactin sequence (hPrl), extending from amino acids 90 to 129 is shown on the upper line, while a portion of human growth hormone (hGH) extending from amino acids 85 to 120 is shown below. Amino acids 90–94 of prolactin are closely homologous with amino acids 85–89 of growth hormone, and amino acids 124–129 of prolactin are homologous with amino acids 115–120 of growth hormone; thus, a gap four amino acids long must be introduced in the growth hormone sequence somewhere between residues 89 and 115. If this is done by introducing only one gap, then there are two possible pairings for each amino acid; these pairings are shown by the lines. The possible pairings are assigned scores based on the "acceptability" of an amino acid replacement, where a score of zero corresponds to random chance (Dayhoff *et al.*, 1978). The scores are subtracted and the net difference is assigned to the more probable of the two pairs. If the vertical alignment is favored the acceptability score is indicated just below the prolactin sequence; if the diagonal alignment is favored, the score is indicated just above the growth hormone sequence. For example, leucine No. 93 of hGH could be aligned with either leucine No. 98 of hPrl or valine No. 102 of hPrl. The Dayhoff table assigns a score of +6 to a leucine–leucine change and a score of +2 to a leucine–valine change, so we assign a net score of 4 to the leucine–leucine alignment. The next amino acid in hGH, arginine No. 94, could thus be aligned with either valine No. 99 or arginine No. 103 in hPrl. An arginine–valine change gets a score of −2 while an arginine–arginine change is scored as +6, so that we assign the arginine–arginine alignment a net score of +8 and place the gap in hGH between the amino acids 93 and 94, whereas the computer program put it between amino acids 89 and 90. Both visual inspection and mathematical summing in all frames assign the gap to the position shown, between amino acids 93 and 94 of human growth hormone.

acid pair distances based on comparisons of amino acid volume and polarity (Miyata *et al.,* 1979). Our procedure for aligning sequences is shown in Fig. 2. Using this procedure, our alignments of human growth hormone and prolactin and of rat growth hormone and prolactin result in aligning the sites where intervening sequences occur in the native genes, suggesting the alignments are valid. The various alignments are shown in Fig. 3.

Comparison of the amino acid sequences of GH and Prl in cattle, rats, and humans give consistent results. Using the alignments shown in Fig. 1, we count the number of amino acids that are identical or which are replaced. The number of amino acids that are different between GH and Prl range from 79.6% in humans to 76.5% in cattle. Using a UEP figure of 4.5, the calculated divergence times are 344–357 million years ago (Miller *et al.,* 1981). When the calculation is made comparing pre-GH to pre-Prl, similar values of about 350 million years are obtained. These calculations suggest that the gene duplication which led to the evolution of separate *GH* and *Prl* genes had to become fixed in a wide variety of species more than 350 million years ago. This is consistent with the observation that all vertebrates have both types of hormones in their pituitaries, and that amphibians diverged from fish about 400 million years ago. Although fish prolactin is obviously not engaged in the control of lactation, it is apparently involved in the maintenance of intracellular osmotic tonicity and control of salt and water flux across membranes, roles that may well be related to the secretion of a hyperosmolar fluid such as milk.

A similar comparison of the sequences of human GH and CS indicates the ancestral genes for these two modern hormones duplicated about 61–64 million years ago. Thus it would appear that GH and Prl diverged in distant evolutionary time, while GH and CS diverged as the result of a much more recent gene duplication. The human *GH* and *CS* genes indeed are closely linked (Selby *et al.,* 1983) on chromosome 17 (Owerbach *et al.,* 1980) while the human *Prl*

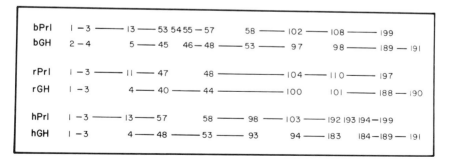

Figure 3.   Alignment of prolactin and growth hormone sequences from bovine (top), rat (middle), and human (bottom) species. The locations of gaps were assigned by the procedure illustrated in Fig. 2.

gene is on chromosome 6 (Owerbach *et al.*, 1981), in good agreement with this hypothesis. However, there is a problem with this simplistic analysis. The paleontologic record is quite detailed and accurate for mammalian speciation, fixing the "mammalian radiation" at 85–100 million years ago. If the evolutionary precursors to various modern mammalian species diverged from one another about 85 million years ago, and the *GH* and *CS* genes duplicated only 60 million years ago, then such a gene duplication would have to have occurred separately in each mammalian species since all placental mammals appear to have a separate GH and CS. This seems highly unlikely.

Another way of using an evolutionary clock is to examine a single gene or protein in a variety of animals. This should tell us when the evolutionary precursors to the modern species diverged from each other, just as comparing related genes in one species tells us when the species diverged. If we compare the amino acid sequences of growth hormone from cattle, rats, and humans and apply the published UEP of 4.0 for GH (Wilson *et al.*, 1977) we would determine that the evolutionary ancestors of humans diverged from the ancestors of cattle and rats 140 million years ago, while the ancestors of cattle and rats diverged 65 million years ago (Miller *et al.*, 1980). Similarly, if we compare prolactin in the same three species using a UEP of 4.5, we would conclude that humans and cattle diverged from rats 170 million years ago and that humans and cattle subsequently diverged from one another 120 million years ago (Miller *et al.*, 1981). While these figures can be changed slightly by using different UEPs or by including (or excluding) the leader peptides of these secreted hormones, the general pattern is always the same: The patterns of mammalian speciation revealed by the GH and Prl sequences are inconsistent with one another, and both are inconsistent with the fossil record (Fig. 4). Thus two different types of inconsistencies are found when the evolutionary clock hypothesis is applied to the growth hormone/prolactin system. First, the calculated time of the GH/CS

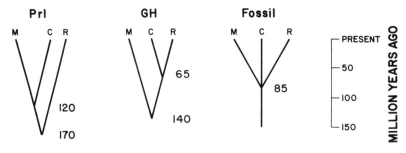

Figure 4. Evolutionary divergence diagrams for man, cattle, and rats as calculated from a comparison of prolactin (left) or growth hormone (center) amino acid sequences assuming a unit evolutionary period of 4.5. Numbers refer to the number of millions of years ago when precursors to the modern species are calculated to have diverged. The pattern of divergence determined by the fossil record is shown at right for comparison.

duplication is unrealistically recent; second, the predicted patterns of mammalian speciation are inconsistent with the well-established fossil record.

It has been suggested that the use of nucleotide sequences for calculating and calibrating evolutionary clocks may be a more accurate and informative procedure for comparing related sequences than the use of amino acid sequences. The availability of nucleotide sequences of cloned cDNAs for GH and Prl from cattle, rats, and humans now permits us to test this hypothesis in the growth hormone/prolactin system. Efstratiadis and colleagues calculated the percent

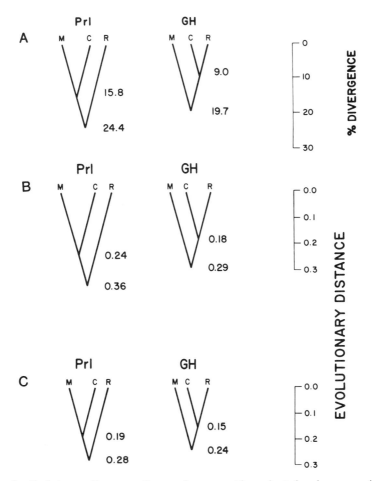

Figure 5. Evolutionary divergence diagrams for man, cattle, and rats based on comparisons of nucleotide sequences using three different mathematical procedures. (A) Percent divergence calculated according to the method of Perler *et al.* (1980). (B) Evolutionary distance calculated according to the three-substitution-type (3ST) model of Kimura (1980). (C) Evolutionary distance calculated according to the two-frequency-class (2FC) model of Kimura (1981).

sequence divergence for nucleotide changes resulting in new amino acids (replacement changes) and for nucleotide changes resulting in the same amino acid (silent changes) (Perler *et al.*, 1980; Efstratiadis *et al.*, 1980). This system is based on the "selectionist" Darwinian theory of molecular evolution, which states that nucleotide replacements are "fixed" in the evolution of genes by positive environmental selection of the altered amino acid sequence (Clarke, 1970; Richmond, 1970; Blundell and Wood, 1975). An alternative hypothesis, the "neutral" theory of molecular evolution, argues that most nucleotide changes and consequent amino acid substitutions that become fixed in evolution are selectively neutral, thus accounting for the many polymorphisms found among proteins with identical functions (Kimura, 1979). Based on this hypothesis, Kimura devised two different ways to compare nucleotide sequences (Kimura, 1980, 1981). We have applied all three of these elaborate mathematical models to the nucleotide sequences for GH and Prl from the three species. The results shown in Fig. 5 are striking: all three procedures give exactly the same patterns determined by the simpler amino acid comparisons.

1.5. Concerted Evolution

Since the evolution of the *GH* and *Prl* genes cannot be described by a simple model, we shall explore how the apparent inconsistencies might be resolved. The evolutionary clock is certainly a useful concept which has been used successfully in several systems. The difficulty with the various ways of applying this concept is that all of the procedures described above deal with the probability of point mutations occurring and becoming fixed in a gene and assume that mutations occur and become fixed at a constant rate in evolutionary time. While point mutation is a major mechanism of evolution, it is not the only one. DNA may be duplicated, rearranged, deleted, or added, either by integration of foreign DNA or by unequal crossover events. When multiple nonallelic genes or pseudogenes exist for a particular peptide the rate of evolutionary divergence can be slowed by "concerted evolution" (Hood *et al.*, 1975; Zimmer *et al.*, 1980; Arnheim *et al.*, 1980; Liebhaber *et al.*, 1981). In concerted evolution, unequal crossover events lead to polygenic states, permitting the elimination of mutant genes while normal ones continue to be expressed, thus slowing the rate of fixation of mutations in a population. Concerted evolution is favored by gene configurations favoring unequal crossovers, e.g., the presence of multiple nonallelic genes and the presence of long, homologous flanking regions or intervening sequences. Thus the occurrence of a gene duplication could increase the opportunities for concerted evolution, thereby slowing the rate of fixation of point mutations.

Human growth hormone and chorionic somatomammotropin exhibit the hallmark of concerted evolution (Zimmer *et al.*, 1980) in that the two peptides are more alike within the human species than is hGH with the growth hormone in various other species. Human growth hormone and chorionic somatomammotropin have 80% amino acid and 92% nucleotide homology (Martial *et al.*, 1979) whereas human and rat or bovine growth hormone have 64–67% amino acid and 75–77% nucleotide homologies, respectively (Miller *et al.*, 1980). The genes for human growth hormone and chorionic somatomammotropin are closely linked on chromosome 17 (Owerbach *et al.*, 1980) and both genes exist in multiple copies (Selby *et al.*, 1983). Furthermore, the mRNA for hCS shares several features with the mRNAs for human, bovine, and rat growth hormone: (1) it has a high G and C content in codon third positions; (2) it has a UAG termination codon; (3) its 3' untranslated region is homologous with the 3' untranslated regions in the three growth hormone mRNAs; and (4) it is more homologous with each of the three growth hormone mRNAs than it is with human prolactin mRNA. Thus these observations suggest that the human chorionic somatomammotropin gene is a duplicated, slightly variant growth hormone gene that has evolved by concerted evolution.

Concerted evolution may not be so important in the evolution of prolactin genes. The human prolactin gene probably cannot interact with growth hormone genes, being segregated on chromosome 6 (Owerbach *et al.*, 1981). Furthermore, analysis of a rat genomic DNA library reveals no evidence of multiple rat prolactin genes, but does suggest that the intervening sequences in the rat prolactin gene are much larger than in the growth hormone genes (Gubbins *et al.*, 1980; Cooke and Baxter, 1982). Thus it is possible that the prolactin and growth hormone genes are evolving by several different mechanisms, and that the dominant mechanism may differ between the two hormones and among various species. These observations imply that evolutionary analysis of related protein and gene sequences may be less straightforward than one might initially expect.

1.6. Structure of Native Genes

In addition to the seven known cDNA structures (Miller *et al.*, 1980, 1981, 1982; Martial *et al.*, 1979; Seeburg *et al.*, 1977; Sasavage *et al.*, 1982; Cooke *et al.*, 1980, 1981; Gubbins *et al.*, 1980; Shine *et al.*, 1977; Goodman *et al.*, 1980), the structures of the native genes for rat (Barta *et al.*, 1981) and human (DeNoto *et al.*, 1981) GH, rat prolactin (Gubbins *et al.*, 1980; Cooke and Baxter, 1982), and human CS (Fiddes *et al.*, 1979; Selby *et al.*, 1983) are now known. In general, gene structure is organized into four regions, the 5' flanking region, the sequences which encode mRNA, the intervening sequences, and the 3' flanking region. The 5' flanking regions lie proximal to the coding sequences, and contain DNA sequences to which the RNA polymerase must bind (promoters), and other regulatory elements, such as those influenced by the hormones that regulate *GH* gene transcription. The presumed promoter sequence in eukaryotic genes is the so-called Goldberg–Hogness box, a segment of DNA having the sequence TATAA, generally located about 31 bases proximal to the site where transcription is initiated (Goldberg, 1979). Another region, generally having the sequence CAAT, is found in most eukaryotic genes approximately 40 bases proximal to the TATAA sequence (Benoist *et al.*, 1980). These two regions are highly homologous among the various *GH* and *Prl* genes, but other portions of the 5' untranslated regions show less homology, the degree of which is generally proportional to the degree of homology in the coding sequences. The homologies of the coding sequences are essentially as described for the cDNAs (above) and will not be discussed further.

Intervening sequences are stretches of DNA which interrupt coding sequences, but do not themselves encode part of a mature mRNA molecule. Intervening sequences (sometimes termed "introns") and coding sequences are initially transcribed into a precursor to mRNA. The pre-mRNA is then processed within the nucleus to remove the RNA corresponding to the intervening sequences, yielding mature mRNA which is then translocated to the cytoplasm for translation. No function has yet been found for intervening sequences or their transcripts. It has been argued that intervening sequences delineate the boundaries of primitive gene precursor components, and may thus delineate various "functional domains" of the encoded proteins (Gilbert, 1978). The intervening sequences in immunoglobulin and hemoglobin genes do indeed appear to divide these proteins into "domains" having different biological activities. However, the lack of intervening sequences between the various, clearly defined domains of proopiomelanocortin (the precursor to corticotropin, endorphin, and other hormones) indicates this hypothesis has limits (Whitfield *et al.*, 1982).

Although intervening sequences have no known function, their presence makes genes longer, thus they increase the chances of a recombinational event occurring within a gene. This could permit the reassortment of various coding sequences (and hence, functional domains) in related genes. Such mechanisms are likely to be involved in concerted evolution. The genes for GH, Prl, and CS each contain four intervening sequences located at precisely homologous positions within the coding sequences. The homology of these locations, despite great differences in the length of the intervening sequences, is further strong support for the hypothesis that these genes arose from a common evolutionary ancestor.

2. Regulation

2.1. Background

Growth hormone production in the mammalian pituitary may be hormonally regulated (for review, see Zachman and Prader, 1972). Confusion arises in a review of the literature because it appears that laboratory rats and human subjects may employ different regulatory systems. Additional confusion arises if cell culture experiments, which are divorced from other regulatory influences (e.g., the hypothalamus) are simplistically interpreted as reflecting *in vivo* physiology. Despite these difficulties, various lines of evidence indicate a major role for glucocorticoid and thyroid hormones in the regulation of GH synthesis. Supraphysiological plasma concentrations of glucocorticoid hormones are associated with poor growth in human beings (Baxter, 1978; Loeb, 1976; Daughaday *et al.*, 1975; Ballard, 1979; Strickland, 1972) and rats (Bellamy, 1964; Winick and Coscia, 1968; Sawano *et al.*, 1969). Excess glucocorticoids appear to increase the GH content of rat pituitaries (Lewis *et al.*, 1965) but decrease GH content in human pituitaries (Suda *et al.*, 1980), as well as decreasing plasma GH concentrations (Tyrrell *et al.*, 1977). Physiological concentrations of glucocorticoids increase GH production by cultured rat, monkey, or human pituitary cells *in vitro* (Kohler *et al.*, 1968, 1969; Bridson and Kohler, 1970) and supraphysiological concentrations of glucocorticoids increase GH production by cultured rat pituitary tumor cells even further (Tashjian *et al.*, 1968; Dannies and Tashjian, 1973; Bancroft *et al.*, 1969; Martial *et al.*, 1977a, 1977b). Thus most laboratory data indicate a glucocorticoid-mediated enhancement of GH in the rat, while extensive clinical data indicate the contrary in humans.

Thyroid hormone may also play a major role in growth hormone physiology. Hyper- and hypothyroidism appear to be directly correlated, respectively, with increased or decreased pituitary GH content (Peake *et al.*, 1973; Hervas *et al.*, 1975; Surks and DeFesi, 1977; DeFesi *et al.*, 1979) and a causal relationship was suggested by the restoration of pituitary GH content by treatment of hypothyroid rats with thyroxine (Lewis *et al.*, 1965). However, human patients with hyper- or hypothyroidism have variable responses of plasma GH in response to provocative stimuli; furthermore, the degree of GH response does not correlate with the degree of hypo- or hyperthyroidism (Katz *et al.*, 1969). Therefore the studies in cultured rat pituitary tumor cells should be viewed as a model for hormonal regulation of gene expression, but should not necessarily be construed to be relevant to human physiology.

2.2. Use of Cultured Rat Pituitary Cells

Many of these hormonal effects are preserved in cultured rat pituitary tumor cells (GH cells) (Tashjian *et al.*, 1968; Dannies and Tashjian, 1973). These cells secrete either growth hormone or prolactin (or both) into the culture medium. Growth hormone production increases in response to glucocorticoids, and can also be modulated by other stimuli (Dannies and Tashjian, 1973; Bancroft *et al.*, 1969). By continuous labeling and by pulse–chase experiments, Yu and Bancroft (1977) showed that newly synthesized growth hormone is secreted after a lag of only 15 min. They also showed that glucocorticoids do not influence the secretion or intracellular degradation of growth hormone, but act directly on its synthesis. Therefore, by measuring the hormone released into the medium, it was possible to estimate precisely the amount of growth hormone synthesis in the absence of complications arising from secretion or degradation.

Glucocorticoid-mediated increases in growth hormone production occur in cells cultured in serum containing thyroid hormone (3,5,3' triiodo-L-thyronine, T_3) (Bancroft et al., 1969; Martial et al., 1977a; Seo et al., 1977; Shapiro et al., 1978) but not in cells grown in serum lacking thyroid hormone (Martial et al., 1977b; Samuels et al., 1977). In contrast, the thyroid hormone response can be observed in cells grown in the absence or presence of glucocorticoids (Samuels et al., 1978). When glucocorticoid and thyroid hormones are given together, production of growth hormone is greater than with either T_3 or glucocorticoid alone (Martial et al., 1977b; Samuels et al., 1977, 1978). This synergistic effect suggests that the two classes of hormones act independently to regulate the production of growth hormone. Ivarie et al. (1981) examined the response of GH cells to glucocorticoid and thyroid hormones by studying the rate of synthesis of various proteins. They showed that either thyroid hormone or glucocorticoids alone increase growth hormone synthesis, and that the effect of both hormones was synergistic. In addition, they showed that the effects of these hormones were independent, since the synthesis of some proteins was affected by T_3, while others were affected by glucocorticoids. Therefore, rat pituitary cell lines are a useful model for studying the mechanisms by which glucocorticoid and thyroid hormones control the expression of the growth hormone gene.

The production of growth hormone entails at least six events. First, the gene for GH must be transcribed into RNA; second, this transcript (pre-mRNA) must be processed within the nucleus into mature mRNA; third, the mRNA must be translocated to the cytoplasm; fourth, the mRNA must attach to ribosomes and be translated into pre-GH; fifth, the pre-GH must enter the endoplasmic reticulum where it is processed to mature GH and packaged; sixth, it must be secreted from the normal cell on demand. Since cultured GH cells do not store and secrete GH in response to physiological stimuli, they form a simpler system for studying GH synthesis, since their release of GH is constituitive.

2.3. Effects of Thyroid and Glucocorticoid Hormones on GH Production in Cultured Cells

Samuels and Shapiro (1976) showed that the thyroid hormone effect is blocked when RNA synthesis is blocked; however when protein synthesis is blocked, an intermediate necessary for stimulation of growth hormone synthesis accumulates in response to thyroid hormone. Although these findings were indirect, they suggested that thyroid hormone may regulate production of growth hormone mRNA. To test this hypothesis, Seo et al. (1977) isolated RNA from cultured rat pituitary tumor cells (GH_3 subline) and translated it in a cell-free rabbit reticulocyte lysate system. They showed that the T_3-stimulated growth hormone synthesis was indeed accompanied by a specific increase in growth hormone mRNA activity. Martial *et al.* (1977a, 1977b) measured the effects of both glucocorticoid and thyroid hormones on growth hormone mRNA. They showed that T_3 and the synthetic glucocorticoid dexamethasone (Dex) stimulated increases in growth hormone production by 2.5- and 3.8-fold, respectively. In addition, there were comparable increases in the capacity of RNA from hormone-treated cells to direct synthesis of growth hormone in a cell-free wheat germ translation system, suggesting the hormones either increased the amount of GH mRNA or increased its translational efficiency, or both. Wegnez *et al.* (1982) measured the effect of hormonal treatments on the translational efficiency of GH mRNA molecules in GH_3 cells. Their data demonstrated that T_3 had no influence on the translational efficiency of GH mRNA, while dexamethasone could either decrease translational efficiency (in serum-substitute medium) or increase it (in serum-containing medium). However, these influences are not specific for GH mRNA, but appear to be a general effect on all mRNAs. Thus, the hormone-induced increases in translatable cytoplasmic GH mRNAs seem to arise from an increase in cytoplasmic GH mRNA sequences. However, the studies mentioned so far all measured hormonal effects on GH mRNA concentrations by indirect methods.

2.4. Hormonal Effect on GH mRNA Content

A direct method for measuring the concentration of a specific mRNA is by RNA/DNA hybridizations in solution. This technique requires a very pure, radioactively labeled probe, i.e., DNA sequences complementary to growth hormone mRNA (cDNA), to distinguish growth hormone mRNA sequences from other mRNAs in low abundance. In one study, Martial *et al.* (1977a) used a cDNA probe prepared from partially purified GH mRNA, and showed that GH mRNA copies do parallel the mRNA activity assessed in cell-free translation assays. However, these experiments were not performed with a pure cDNA probe and could only estimate the amount of GH mRNA, not their synthesis. Pure probes became available only after cloning the cDNA to rat GH mRNA (Seeburg *et al.*, 1977).

Wegnez *et al.* (1982) used a pure radioactive probe in RNA/DNA hybridization experiments to estimate the glucocorticoid and thyroid hormone influences on GH mRNA copies in cultured rat pituitary tumor cells (GH_3D_6 subline). They were able to overcome some problems arising from commonly used procedures for radioactive labeling of the DNA (i.e., formation of hairpin loops which may occur during nick-translation and result in high background levels of hybridization) by labeling only the DNA strand of the rat GH that hybridizes to rat GH mRNA (the cDNA strand) (O'Farrell, 1981). In addition, they were able to overcome the problem of endogenous hormones in the medium in which the cells were grown. In cells cultured in the presence of 10% calf serum, growth hormone production increases only minimally in response to added T_3, presumably because this hormone is already present at physiological concentrations in the serum (Martial *et al.*, 1977a). In addition, in this medium, dexamethasone significantly increases growth hormone production. Therefore, to bypass the problem of endogenous T_3 in calf serum, Wegnez *et al.* used two other media: one contained serum from a thyroidectomized calf (hypothyroid medium) which was depleted of thyroid hormones (Samuels and Tsai, 1973). In this case, T_3 induces growth hormone production greater than 10-fold, and dexamethasone has a much smaller influence (Martial *et al.*, 1977b; Samuels *et al.*, 1977, 1979). The second medium contained a chemically defined serum substitute preparation (Bauer *et al.*, 1976) and therefore contained no endogenous T_3 or glucocorticoid. In addition, other hormones and factors present in serum may affect growth hormone regulation, but are not present in serum-substitute, permitting one to study the effects of glucocorticoid and thyroid hormones on growth hormone regulation less ambiguously.

When the rat pituitary tumor cells (GH_3D_6 subline) were grown in serum substitute medium and stimulated for 72 hr with either dexamethasone or T_3, cytoplasmic GH mRNA content increased four- and sixfold, respectively (Table II) (from < 2 to 8 and from < 2 to 12 molecules of GH mRNA/cell). When added together, the two classes of hormones act synergistically, causing a 19-fold increase in GH mRNA content (from < 2 to 38 molecules of GH mRNA/cell). When similar experiments were performed with hypothyroid medium, dexamethasone increased cytoplasmic GH mRNA fourfold (Table II) (from < 5 to 20 molecules of GH mRNA/cell), T_3 increased GH mRNA 26-fold (to 130 molecules of GH mRNA/cell), and both hormones together increased GH mRNA 66-fold (to 330 molecules of GH mRNA/cell). Thus, while the results with the two different media were qualitatively similar, they were not quantitatively identical, suggesting that the hypothyroid medium contains other factors not present in

Table II. Hormonal Effects on Growth Hormone mRNA
Copies in Cultured Rat Pituitary Cells[a]

	Molecules of growth hormone mRNA/cell		
	GH_3D_6 cells		GC cells
Hormonal treatment	Serum-substitute medium	Hypothyroid medium	Hypothyroid medium
Control	<2	<5	100
+ Dex	8	20	450
+ T_3	12	130	800
+ Dex + T_3	38	330	2000

[a] GH_3D_6 cell data are from Wegnez *et al.* (1982); GC cell data are from Spindler *et al.* (1982).

serum substitute medium which modulate either GH mRNA synthesis or GH mRNA half-life.

Other experiments measuring cytoplasmic GH mRNA content were performed with another subline of rat pituitary tumor cells, GC cells, (Spindler *et al.,* 1982). GC cells were grown in hypothyroid medium and then induced for 72 hr with dexamethasone with or without T_3. The number of GH mRNA copies is increased approximately 4.5-fold by dexamethasone (Table II) (from 100 to 450 molecules of GH mRNA/cell), eightfold by T_3 (from 100 to 800 molecules of GH mRNA/cell), and about 20-fold (from 100 to 2000 molecules of GH mRNA/cell) by both dexamethasone and T_3. Comparing these data to those of Wegnez *et al.,* dexamethasone and T_3 again exert a qualitatively similar effect on GH mRNA copy number, but quantitative differences may arise from the difference in the cell sublines used in each experiment. Nevertheless, the conclusions that can be drawn are that the effect of glucocorticoid and thyroid hormones on GH production may be explained by increasing the amount of cytoplasmic GH mRNA. These results, however, do not distinguish between an increase in GH mRNA synthesis and an increase in GH mRNA stability (i.e., an increase in GH mRNA half life).

2.5. Hormonal Effects on a GH mRNA Synthesis

To determine whether this increase in cytoplasmic GH mRNA is due to an increase in its synthesis, Dobner *et al.* (1981) and Cathala *et al.* (1983) examined the effects of glucocorticoids and T_3 on growth hormone pre-mRNA by Northern blot analysis (Thomas, 1980). This technique will give only a qualitative estimate of the synthesis of the primary GH transcript (pre-mRNA) and processing of this primary transcript into a mature GH mRNA. Both groups found that GH pre-mRNA was increased in GH_3 cells following glucocorticoid and/or thyroid hormone treatment. Dobner *et al.* detected a 2.7 kb nuclear GH-specific RNA sequence, (GH pre-mRNA) and a 1.0 kb cytoplasmic species of GH mRNA in GH_3 cells. However, because the amount of these RNAs in control nuclei were too low to measure, they were not able to quantitate the hormonal induction of these mRNAs. Nevertheless, it could be concluded that part of the influence of glucocorticoid and thyroid hormones is to increase the synthesis of GH pre-mRNA. These results do not, however, rule out effects on mRNA processing and/or stability.

To determine if the effects of glucocorticoid and thyroid hormones on increasing GH pre-mRNA are due to an increase in the synthesis of this molecule, it is necessary to measure its rate of synthesis, i.e., the rate of transcription of the *GH* gene under different hormonal stimuli. We have quantitated the relative activity of the *GH* gene in GC cells. Nuclei from hormone-treated and control cells were isolated and incubated with radiolabeled ribonucleoside triphosphates under conditions in which the endogenously bound RNA polymerase molecules

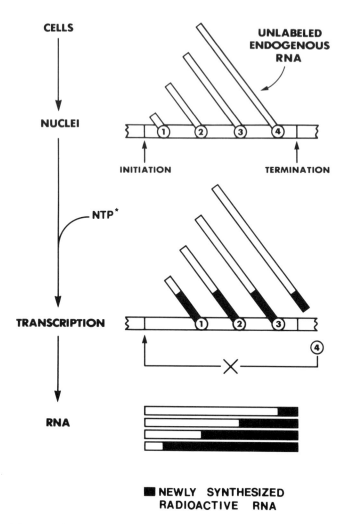

■ NEWLY SYNTHESIZED RADIOACTIVE RNA

Figure 6.   Diagrammatic representation of the cell-free transcription system used to estimate the RNA polymerases initiated *in vivo*. Nuclei were isolated and incubated in the presence of radioactive ribonucleoside triphosphates (NTP*). The nuclei contain the *GH* gene with molecules of RNA polymerase II which had been attached *in vivo*, thus initiating transcription. Transcription proceeds *in vitro*, resulting in the incorporation of radioactive nucleoside triphosphates into newly elongated RNA molecules (dark bars). The newly synthesized RNA is thus radioactive and those molecules containing growth hormone mRNA sequences can be detected by hybridization with cloned cDNA to growth hormone mRNA.

can proceed with transcription and chain termination, but do not reinitiate RNA synthesis (Zylber and Penman, 1971; Reeder and Roeder, 1972) (Fig. 6). Since the newly synthesized RNAs are labeled in this way, their quantity can be assessed by hybridization to an unlabeled DNA probe. Steps in the preparation of the probes are shown in Fig. 7.

For these studies the GC line of cultured rat pituitary tumor cells was used. The cells were grown to confluency in monolayer in medium containing calf serum, and then maintained for 7 days in medium containing serum from a thyroidectomized calf (hyposerum). This serum contains somewhat variable amounts of cortisol, although usually less than 0.75 μg/dl (Martial *et al.*, 1977b), and frequently little response to exogenously added glucocorticoid can be detected. To study glucocorticoid and thyroid hormone responses, cells were maintained

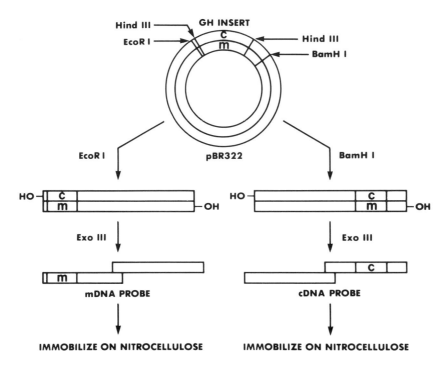

Figure 7.   Preparation of unlabeled growth hormone gene sequences for use as hybridization probes. The plasmid containing rat growth hormone cDNA sequences cloned in the *Hind* III site of pBR322 was cleaved with restriction endonucleases *Bam* HI or *Eco* RI, and the linear DNA was digested 3' to 5' with Exonuclease III. Cleavage with *Bam* HI followed by Exonuclease III digestion exposes the DNA strand which is normally transcribed into RNA (labeled "C", as it is the strand which is complementary to the mRNA). Cleavage with *Eco* RI followed by Exonuclease III exposes the nontranscribed strand (labeled "M," as its sequence is equivalent to an mRNA molecule). These DNAs were then immobilized onto nitrocellulose filters for hybridization.

in medium containing hyposerum and then induced for 4 or 72 hr with dexamethasone (1 µM), T_3, (10 nM), or a combination of Dex and T_3. Cells were then harvested and nuclei prepared by gentle homogenization of the cells with the nonionic detergent NP40. Nuclei were separated from the remainder of the cellular contents by brief centrifugation. The nuclear pellet was then incubated in the presence of [^{32}P]ribonucleoside triphosphates, allowing the endogenously initiated RNA polymerase to elongate RNA chains. The radiolabeled RNA products were then purified by DNAase and sodium dodecyl sulfate (SDS)-protease K digestion and phenol extraction and unincorporated nucleotide triphosphates were removed by filtration (Evans *et al.*, 1981). The radiolabeled RNA was hybridized either to nitrocellulose filters containing cloned cDNA to rat GH mRNA or to the complement of the cDNA (GH mDNA, probe; Fig. 7). The filters were then washed extensively and the specifically bound RNA was quantitated by measuring the radioactivity on the filters (McKnight and Palmiter, 1979). The results from such an experiment are shown in Fig. 8.

After either 4 or 72 hr of hormonal stimulation, glucocorticoids only slightly increased the ability of nuclei to incorporate radioactivity into material that hybridizes specifically with the rGH cDNA probe. After 72 hr of hormonal stimulation, T_3 increases GH gene transcription about fivefold. There does not appear to be any synergism between T_3 and glucocorticoids after 72 hr stimulation.

After 4 hr of hormonal stimulation, T_3 greatly increases *GH* gene transcription

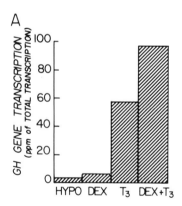 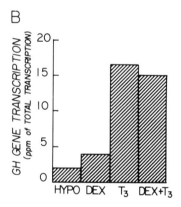

Figure 8. Transcription of the growth hormone gene in isolated nuclei. Cultured rat pituitary cells (GC subline) were grown to confluency in monolayer in media containing 10% calf serum, and then maintained for 7 days in media containing 10% thyroidectomized calf serum (hyposerum). Cells were induced with dexamethasone (Dex) (1 µM), 3,5,3′triiodo-L-thyronine (T_3) (10 nM), or both hormones for 4 hr (A) or 72 hr (B). Nuclei were then isolated and incubated in the presence of [^{32}P]ribonucleoside triphosphates. The [^{32}P]-labeled RNA was isolated and hybridized to filters containing the cloned rat growth hormone cDNA strand. Values are represented as cpm of [^{32}P]RNA hybridized to the rat growth hormone cDNA/million cpm in the hybridization reaction (ppm).

(about 16-fold) and together, Dex and T_3 act synergistically, causing a 28-fold increase in gene transcription. In addition, the effects of thyroid and glucocorticoid hormones on *GH* gene transcription are abolished if the experiments are performed in the presence of 2 μg/ml α-amanitin. These results imply that RNA polymerase II is the enzyme responsible for transcribing the *GH* gene (Weinmann and Roeder, 1974; Weil and Blatti, 1974).

Transcription of the *GH* gene is from one strand of DNA only. There were only background levels of hybridization with the mDNA probe, suggesting that the "antisense" strand of the rat *GH* gene is not transcribed. These data therefore suggest that glucocorticoid and thyroid hormones increase the number of RNA polymerase II molecules engaged in transcribing the growth hormone gene. Given that the hormones also increased growth hormone pre-mRNA, the results further imply that these two classes of hormones increase the rate of transcription of the gene and that this increase is at the level of initiation of transcription. Since the increase in transcription can be readily observed by 4 hr, these data suggest that the effects of T_3 and Dex are rather rapid. Therefore, the hormonal influence may be direct rather than through the induction of some other protein(s).

To determine if prior synthesis of other proteins is required for initiating *GH* gene transcription, a series of experiments was performed in the presence of 0.1 mM cycloheximide, which inhibits translation but not transcription. The results demonstrate that cycloheximide does not affect either the T_3- or dexamethasone-induced increase in *GH* gene transcription, impling that both T_3 and dexamethasone act directly on chromatin to increase *GH* gene transcription.

Similar results were obtained *in vivo* in rats, using thyroidectomized–adrenalectomized rats (stimulating "hypo"-cellular conditions) injected with dexamethasone, T_3, or both hormones (Mellon *et al.*, 1983). Under these experimental conditions, glucocorticoids and thyroid hormone markedly increased pituitary growth hormone gene transcription.

It is of interest to compare the relative changes in growth hormone gene transcription assessed by the experiments described above, with the changes in growth hormone mRNA. In the absence of T_3, dexamethasone induced a two- to threefold increase in growth hormone gene transcription at 4 or 72 hr, whereas at 72 hr the mRNA was increased sevenfold. This result may imply that the effects on transcription do not totally explain the glucocorticoid influence on growth hormone mRNA. Thus, it is possible that the steroid–receptor complex (directly or indirectly) also affects mRNA stability. It should be noted that in another glucocorticoid-responsive system in which such influence for RNA stability has been assessed (mouse mammary tumor virus), it was proposed that the steroid actually decreases rather than increases mRNA stability (Ringold *et al.*, 1977).

ACKNOWLEDGMENTS. Supported by NIH Grant HD 16047 (W. L. M.) and National Research Service Award AM 06588 (S. H. M.). We thank N. L.

Eberhardt for communication of unpublished results, J. D. Baxter for support, and Susan Bromley and Deborah Goodman for typing the manuscript.

References

Arnheim, N., Krystal, M., Schmickel, R., Wilson, G., Ryder, O., and Zimmer, E., 1980, Molecular evidence for genetic exchanges among ribosomal genes on nonhomologous chromosomes in man and apes, *Proc. Natl. Acad. Sci. USA* **77**:7323–7327.

Ballard, P. L., 1979, Glucocorticoids and differentiation, in: *Glucocorticoid Hormone Action* (J. D. Baxter and G. G. Rousseau, eds.), pp. 493–515, Springer-Verlag, New York.

Bancroft, F. C., Levine, L., and Tashjian, A. H., Jr., 1969, Control of growth hormone production by a clonal strain of rat pituitary cells: Stimulation by hydrocortisone, *J. Cell. Biol.* **43**:432–441.

Barta, A., Richards, R. I., Baxter, J. D., and Shine, J., 1981, Primary structure and evolution of the rat growth hormone gene, *Proc. Natl. Acad. Sci. USA* **78**:4867–4871.

Bauer, R. F., Arthur, L. O., and Fine, D. L., 1976, Propagation of mouse mammary tumor cell lines and production of mouse mammary tumor virus in a serum-free medium, *In Vitro* **12**:558–563.

Baxter, J. D., 1978, Mechanisms of glucocorticoid inhibition of growth, *Kidney Int.* **14**:330–333.

Bellamy, D., 1964, Effect of cortisol on growth and food intake in rats, *J. Endocrinol.* **31**:83–84.

Benoist, C., O'Hare, K., Breathnach, R., and Chambon, P., 1980, The ovalbumin gene-sequence of putative control regions, *Nucleic Acids Res.* **8**:127–142.

Blundell, T. I., and Wood, S. P., 1975, Is the evolution of insulin Darwinian or due to selectively neutral mutation? *Nature* **257**:197–203.

Bridson, W. E., and Kohler, P. O., 1970, Cortisol stimulation of growth hormone production by human pituitary tissue in culture, *J. Clin. Endocrinol.* **30**:538–540.

Cathala, G., Savouret, J. F., Martial, J. A., and Baxter, J. D., 1983, Hormonal control of growth hormone pre-mRNA, submitted.

Catt, K. J., Moffat, B., and Niall, H. D., 1967, Human growth hormone and placental lactogen: Structural similarity, *Science* **157**:321.

Clarke, B., 1970, Darwinian evolution of proteins, *Science* **168**:1009–1011.

Cooke, N. E., and Baxter, J. D., 1982, Structural analysis of the prolactin gene suggests a separate origin for its 5′ end. *Nature* **297**:603–606.

Cooke, N. E., Coit, D., Weiner, R. I., Baxter, J. D., and Martial, J. A., 1980, Structure of cloned DNA complementary to rat prolactin messenger RNA, *J. Biol. Chem.* **255**:6502–6510.

Cooke, N. E., Coit, D., Shine, J., Baxter, J. D., and Martial, J. A., 1981, Human prolactin: cDNA structural analysis and evolutionary comparisons, *J. Biol. Chem.* **256**:4006–4016.

Dannies, P. S., and Tashjian, A. H., Jr., 1973, Growth hormone and prolactin from rat pituitary tumor cells, in: *Tissue Culture: Methods and Applications* (P. F. Kruse, Jr. and M. K. Patterson, Jr., eds.), pp. 561–569, Academic Press, New York.

Daughaday, W. J., Herrington, A. C., and Phillips, L. S., 1975, The regulation of growth by endocrines, *Annu. Rev. Physiol.* **37**:211–244.

Dayhoff, M. O., Eck, R. V., and Park, C. M., 1972, A model of evolutionary change in proteins, in: *Atlas of Protein Sequence and Structure,* Vol. 5 (M. O. Dayhoff, ed), pp. 89–99, National Biomedical Research Foundation, Washington, D.C.

Dayhoff, M. O., Schwartz, R. M., and Orcutt, B. C., 1978, A model of evolutionary change in proteins, in: *Atlas of Protein Sequence and Structure,* Vol. 5 Suppl. 3 (M. O. Dayhoff, ed.), pp. 345–352, National Biomedical Research Foundation, Washington, D.C.

DeFesi, C. R., Astier, H. S., and Surks, M. I., 1979, Kinetics of thyrotrophs and somatotrophs during development of hypothyroidism and L-triiodothyronine treatment of hypothyroid rats, *Endocrinology* **104**:1172–1180.

DeNoto, F. M., Moore, D. D., and Goodman, H. M., 1981, Human growth hormone DNA sequence and mRNA structure: Possible alternative splicing, *Nucleic Acids Res.* **9**:3719–3730.

Dobner, P. R., Kawasaki, E. S., Yu, L.-Y., and Bancroft, F. C., 1981, Thyroid or glucocorticoid hormone induces pre-growth hormone mRNA and its probable nuclear precursor in rat pituitary cells, *Proc. Natl. Acad. Sci. USA* **78**:2230–2234.

Efstratiadis, A., Posakony, J. W., Maniatis, T., Lawn, R. M., O'Connell, C., Spritz, R. A., DeRiel, J. K., Forget, B. G., Weissman, S. M., Slightom, J. L., Blechl, A. E., Smithies, O., Baralle, F. E., Shoulders, C. C., and Proudfoot, N. J., 1980, The structure and evolution of the human beta-globin gene family, *Cell* **21**:653–668.

Evans, M. I., Hager, L. J., and McKnight, G. S., 1981, A somatomedin-like peptide hormone is required during the estrogen-mediated induction of ovalbumin gene transcription, *Cell* **25**:187–193.

Fiddes, J. C., Seeburg, P. H., DeNoto, F. M., Hallwell, R. A., Baxter, J. D., and Goodman, H. M., 1979, Structure of genes for human growth hormone and chorionic somatomammotropin, *Proc. Natl. Acad. Sci. USA* **76**:4294–4298.

Gilbert, W., 1978, Why genes in pieces? *Nature* **271**:501.

Goldberg, M., 1979, Sequence analysis of *Drosophila* histone genes, Ph.D. thesis, Stanford University.

Goodman, H. M., DeNoto, F., Fiddes, J. C., Hallewell, R. A., Page, G. S., Smith, S., and Tischer, E., 1980, Structure and evolution of growth hormone-related genes, in: *Mobilization and Reassembly of Genetic Information* (Scott, W. A., Werner, R., Joseph, D. R. and Schultz, J., eds.), pp. 155–179, Academic Press, New York.

Gubbins, E. J., Maurer, R. A., Lagrimini, M., Erwin, C. R., and Donelson, J. E., 1980, Structure of the rat prolactin gene. *J. Biol. Chem.* **255**:8655–8662.

Hervas, F., Morreale de Escobar, G., and Escobar Del Rey, F., 1975, Rapid effects of single small doses of L-thyroxine and triiodo-L-thyronine on growth hormone, as studied in the rat by radioimmunoassay, *Endocrinology* **97**:91–101.

Hood, L., Campbell, J. H., and Elgin, S. C. R., 1975, The organization, expression, and evolution of antibody genes and other multigene families, *Annu. Rev. Genet.* **9**:305–353.

Ivarie, R. D., Baxter, J. D., and Morris, J. A., 1981, Interaction of thyroid and glucocorticoid hormones in rat pituitary tumor cells, *J. Biol. Chem.* **256**:4520–4528.

Katz, H. P., Youlton, R., Kaplan, S. L., and Grumbach, M. M. (1969) Growth and growth hormone III. Growth hormone release in children with primary hypothyroidism and thyrotoxicosis, *J. Clin. Endocrinol. Metab.* **29**:346–351.

Kimura, M., 1979, The neutral theory of molecular evolution, *Sci. Am.* **241**(5):98–126.

Kimura, M., 1980, A simple method for estimating evolutionary rates of base substitutions through comparative studies of nucleotide sequences, *J. Mol. Evol.* **16**:111–120.

Kimura, M., 1981, Estimation of evolutionary distances between homologous nucleotide sequences, *Proc. Natl. Acad. Sci. USA* **78**:454–458.

Kohler, P. O., Bridson, W. E., and Rayford, P. L., 1968, Cortisol stimulation of growth hormone production by monkey adenohypophysis in tissue culture, *Biochem. Biophys. Res. Commun.* **33**:834–840.

Kohler, P. O., Frohman, L. A., Bridson, W. E., Vanha-Perttula, T., and Hammond, J. M., 1969, Cortisol induction of growth hormone synthesis in a clonal line of rat pituitary tumor cells in culture, *Science* **166**:633–634.

Kronenberg, H. M., McDevitt, B. E., Majzoub, J. A., Nathans, J., Sharp, P. A., Potts, J. T., and Rich, A., 1979, Cloning and nucleotide sequence of DNA coding for bovine preproparathyroid hormone, *Proc. Natl. Acad. Sci. USA* **76**:4881–4985.

Lewis, U. J., Cheever, E. V., and Van der Laan, W. P., 1965, Alterations of the proteins of the pituitary gland of the rat by estradiol and cortisol, *Endocrinology* **76**:362–368.

Li, C. H., Dixon, J. S., Lo, T. B., Pankov, Y. M., and Schmidt, K. D., 1967, Amino acid sequence of ovine lactogenic hormones, *Nature* **224**:695–696.

Liebhaber, S. A., Goossens, M., and Kan, Y. W., 1981, Homology and concerted evolution at the alpha-1 and alpha-2 loci of human alpha globin, *Nature* **290**:26–29.

Loeb, J. N., 1976, Corticosteroids and growth, *N. Engl. J. Med.* **295**:547–552.

Martial, J. A., Baxter, J. D., Goodman, H. M., and Seeburg, P. H., 1977a, Regulation of growth hormone messenger RNA by thyroid and glucocortocoid hormones, *Proc. Natl. Acad. Sci. USA* **74**:1816–1820.

Martial, J. A., Seeburg, P. H., Guenzi, D., Goodman, H. M., and Baxter, J. D., 1977b, Regulation of growth hormone gene expression: Synergistic effects of thyroid and glucocorticoid hormones, *Proc. Natl. Acad. Sci. USA* **74**:4293–4295.

Martial, J. A., Hallewell, R. A., Baxter, J. D., and Goodman, H. M., 1979, Human growth hormone: Complementary DNA cloning and expression in bacteria, *Science* **205**:602–607.

McKnight, G. S., and Palmiter, R. D., 1979, Transcriptional regulation of the ovalbumin and conalbumin genes by steroid hormones in chick oviduct, *J. Biol. Chem.* **254**:9050–9058.

Mellon, S. H., Pruss, R. C. M., Baxter, J. D., and Spindler, S. R., 1983, *In vivo* effects of thyroid and glucocorticoid hormones on gene expression: Evidence of transcriptional control of growth hormone gene in thyroidectomized and adrenalectomized rats, Submitted.

Miller, W. L., 1981, Recombinant DNA and the pediatrician, *J. Pediatr.* **99**:1–15.

Miller, W. L., 1982, Bovine prolactin: Corrected cDNA sequence and genetic polymorphisms, *DNA* **1**:313–314.

Miller, W. L., Martial, J. A., and Baxter, J. D., 1980, Molecular cloning of DNA complementary to bovine growth hormone mRNA, *J. Biol. Chem.* **255**:7521–7524.

Miller, W. L., Coit, D., Baxter, J. D., and Martial, J. A., 1981, Cloning of bovine prolactin cDNA and evolutionary implications of its sequence, *DNA* **1**:37–50.

Miyata, T., Miyazawa, S., and Yasunaga, T., 1979, Two types of amino acid substitutions in protein evolution, *J. Mol. Evol.* **12**:219–236.

Nakanishi, S., Inoue, A., Kita, T., Nakamura, M., Chang, A. C. Y., Cohen, S. N., and Numa, S., 1979, Nucleotide sequence of cloned cDNA for bovine corticotropin-beta-lipotropin, *Nature* **278**:423–427.

Niall, H. D., Hogan, M. L., Sayer, R., Rosenblum, I. Y., and Greenwood, F. C., 1971, Sequences of pituitary and placental lactogenic and growth hormones: Evolution from a primordial peptide by gene duplication, *Proc. Natl. Acad. Sci. USA* **68**:866–869.

O'Farrell, P., 1981, Replacement synthesis method of labeling DNA fragments, *Focus* **3**:1–3.

Owerbach, D., Rutter, W. J., Martial, J. A., Baxter, J. D., and Shows, T. B., 1980, Genes for growth hormone, chorionic somatomammotropin, and growth hormone-like gene on chromosome 17 in humans, *Science* **209**:289–292.

Owerbach, D., Rutter, W. J., Cooke, N. E., Martial, J. A., and Shows, T. B., 1981, The prolactin gene is located on chromosome 6 in humans, *Science* **212**:815–816.

Peake, G. T., Birge, C. A., and Daughaday, W. H., 1973, Alterations of radioimmunoassayable growth hormone and prolactin during hypothyroidism, *Endocrinology* **92**:487–493.

Perler, F., Efstratiadis, A., Lomedico, P., Gilbert, W., Kolodner, R., and Dodgson, J., 1980, The evolution of genes: The chicken proinsulin gene. *Cell* **20**:555–566.

Proudfoot, N. J., and Brownlee, G. G., 1976, 3′ Non-coding region sequences in eukaryotic messenger RNA, *Nature* **263**:211–214.

Reeder, R. H., and Roeder, R. G., 1972, Ribosomal RNA synthesis in isolated nuclei, *J. Mol. Biol.* **67**:433–441.

Richmond, R. C., 1970, Non-Darwinian evolution: A critique. *Nature* **225**:1025–1028.

Ringold, G. M., Yamamoto, K. R., Bishop, J. M., and Varmus, H. E., 1977, Glucocorticoid-stimulated accumulation of mouse mammary tumor virus RNA: Increased rate of synthesis of viral RNA. *Proc. Natl. Acad. Sci. USA* **74**:2879–2883.

Samuels, H. H., and Shapiro, L. E., 1976, Thyroid hormone stimulates *de novo* growth hormone synthesis in cultured GH$_1$ cells: Evidence for the accumulation of a rate limiting RNA species in the induction process, *Proc. Natl. Acad. Sci. USA* **73**:3364–3373.

Samuels, H. H., and Tsai, J. S., 1973, Thyroid hormone action in cell culture: Demonstration of nuclear receptors in intact cells and isolated nuclei, *Proc. Natl. Acad. Sci. USA* **70:**3488–3492.

Samuels, H. H., Horwitz, Z. D., Stanley, F., Casanova, J., and Shapiro, L. E., 1977, Thyroid hormone controls glucocorticoid action in cultured GH₁ cells, *Nature* **268:**254–256.

Samuels, H. H., Klein, D., Stanley, F., and Casanova, J., 1978, Evidence for thyroid hormone-dependent and independent glucocorticoid actions in cultured cells, *J. Biol. Chem.* **253:**5895–5898.

Samuels, H. H., Stanley, F., and Shapiro, L. E., 1979, Control of growth hormone synthesis in cultured GH₁ cells by 3,5,3′-triiodo-L-thyronine and glucocorticoid agonists and antagonists: Studies on the independent and synergistic regulation of the growth hormone response, *Biochemistry* **18:**715–721.

Sasavage, N. L., Nilson, J. H., Horowitz, S., and Rottman, F. M., 1982, Nucleotide sequence of bovine prolactin messenger RNA—evidence for sequence polymorphism, *J. Biol. Chem.* **257:**678–681.

Sawano, S., Arimura, A., Schally, A. V., Reding, T. W., and Schapiro, S., 1969, Neonatal corticoid administration: Effects upon adult pituitary growth hormone and hypothalamic growth hormone-releasing hormone, *Acta Endocrinol.* **61:**57–67.

Seeburg, P. H., Shine, J., Martial, J. A., Baxter, J. D., and Goodman, H. M., 1977, Nucleotide sequence and amplification in bacteria of the structural gene for rat growth hormone, *Nature* **270:**486–494.

Selby, M., Barta, A., Birnbaum, M., Baxter, J. D., Bell, G. I., and Eberhardt, N. L., Structure and linkage of the human chorionic somatomammotropin gene family, Submitted.

Seo, H., Vassart, G., Brocas, H., and Refetoff, S., 1977, Triiodothyronine stimulates specifically growth hormone mRNA in rat pituitary tumor cells, *Proc. Natl. Acad. Sci. USA* **74:**2054–2058.

Shapiro, L. E., Samuels, H. H., and Yaffe, B. M., 1978, Thyroid and glucocorticoid hormones synergistically control growth hormone mRNA in cultured GH₁ cells, *Proc. Natl. Acad. Sci. USA* **75:**45–49.

Sherwood, L. M., 1967, Similarities in the chemical structure of human placental lactogen and pituitary growth hormone, *Proc. Natl. Acad. Sci. USA* **58:**2307–2314.

Shine, J., Seeburg, P. H., Martial, J., Baxter, J. D., and Goodman, H. M., 1977, Construction and analysis of recombinant DNA for human chorionic somatomammotropin, *Nature* **270:**494–499.

Spindler, S. R., Mellon, S. H., and Baxter, J. D., 1982, Growth hormone gene transcription is regulated by thyroid and glucocorticoid hormones in cultured rat pituitary tumor cells, *J. Biol. Chem.* **257:**11627–11632.

Strickland, A. L., 1972, Growth retardation in Cushing's syndrome, *Am. J. Dis. Child.* **124:**207–213.

Suda, T., Demura, H., Demura, R., Jibiki, K., Tozawa, F., and Shizume, K., 1980, Anterior pituitary hormones in plasma and pituitaries from patients with Cushing's disease, *J. Clin. Endocrinol. Metab.* **51:**1048–1053.

Surks, M. I., and DeFesi, C. R., 1977, Determination of the cell number of each cell type in the anterior pituitary of euthyroid and hypothyroid rats, *Endocrinology* **101:**946–958.

Tashjian, A. H., Jr., Yasumura, Y., Levine, L., Sato, G. H., and Parker, M. L., 1968, Establishment of clonal strains of rat pituitary tumor cells that secrete growth hormone, *Endocrinology* **82:**342–352.

Thomas, P. S., 1980, Hybridization of denatured RNA and small DNA fragments transferred to nitrocellulose, *Proc. Natl. Acad. Sci. USA* **77:**5201–5205.

Tyrrell, J. B., Wiener-Kronish, J., Lorenzi, M., Brooks, R. M., and Forsham, P. H., 1977, Cushing's disease: Growth hormone response to hypoglycemia after correction of hypercortisolism, *J. Clin. Endocrinol. Metab.* **44:**218–221.

Wegnez, N., Schachter, B. S., Baxter, J. D., and Martial, J. A., 1982, Hormonal regulation of growth hormone mRNA, *DNA* **1:**145–153.

Weil, P. A., and Blatti, S. P., 1974, Partial purification and properties of calf thymus deoxyribonucleic acid dependent RNA polymerase III, *Biochemistry* **14:**1636–1642.

Weinmann, R., and Roeder, R. G., 1974, Role of DNA-dependent RNA polymerase III in the transcription of the tRNA and 5S RNA genes, *Proc. Natl. Acad. Sci. USA* **71:**1790–1794.

Whitfeld, P. L., Seeburg, P. H., and Shine, J., 1982, The human pro-opiomelanocortin gene: Organization, sequence, and interspersion with repetitive DNA, *DNA* **1:**133–143.

Wilson, A. C., Carlson, S. S., and White, T. J., 1977, Biochemical evolution, *Annu. Rev. Biochem.* **46:**573–639.

Winick, M., and Coscia, A., 1968, Cortisone induced growth failure in neonatal rats, *Pediatr. Res.* **2:**451–455.

Yu, L.-Y., and Bancroft, F. C., 1977, Glucocorticoid induction of growth hormone synthesis in a strain of rat pituitary cells, *J. Biol. Chem.* **252:**3870–3875.

Zachman, M., and Prader, A., 1972, Interactions of growth hormone with other hormones, in: *Human Growth Hormone* (A. S. Mason, ed.), pp. 39–93, Heinemann Medical Books, London.

Zimmer, E. A., Martin, S. L., Beverley, S. M., Kan, Y. W., and Wilson, A. C., 1980, Rapid duplication and loss of genes coding for the alpha-chains of hemoglobin, *Proc. Natl. Acad. Sci. USA* **77:**2158–2162.

Zuckerkandl, E., and Pauling, L., 1962, Molecular disease, evolution and genic heterogeneity, in: *Horizons in Biochemistry* (M. Kasha and B. Pullman, eds.), pp. 189–225, Academic Press, New York.

Zuckerkandl, E., and Pauling, L., 1965, Evolutionary divergence and convergence in proteins, in: *Evolving Genes and Proteins* (V. Bryson and H. J. Vogel, eds.), pp. 97–166, Academic Press, New York.

Zylber, E. A., and Penman, S., 1971, Products of RNA polymerases in HeLa cell nuclei, *Proc. Natl. Acad. Sci. USA* **68:**2861–2865.

10

Androgenic Control of Gene Expression in Rat Ventral Prostate

Malcolm G. Parker

1. Introduction

Spectacular advances in recombinant DNA technology have enabled us to begin to analyze the structure and expression of many eukaryotic genes, including those whose expression is responsive to androgenic steroids. Such studies are essential if we are to understand how steroids influence gene expression because, although we know a great deal about steroid–receptor interactions in the cytoplasm, we know very little about events that take place in the cell nucleus. All classes of steroids appear to stimulate gene expression primarily by stimulating rates of mRNA production (Higgins and Gehring, 1978). It has also been shown, using both biochemical and genetic criteria, that hormone–receptor complexes bind to DNA and that this binding is related to a biological response (Yamamoto *et al.*, 1974; Gronemeyer and Pongs, 1980). In view of this, it is generally postulated that hormone–receptor complexes bind directly to responsive genes, thereby stimulating their rate of transcription. The isolation of such genes is an essential prerequisite for testing this hypothesis. Furthermore, it is hoped that by analyzing the structure and organization of hormone responsive genes we will begin to understand what features are necessary for hormone responsiveness and what distinguishes such genes that give rise to tissue-specific gene expression.

Our studies are concerned mainly with three genes in the rat that are expressed in the ventral prostate under the control of androgenic steroids to produce a

Malcolm G. Parker ● Imperial Cancer Research Fund, P.O. Box 123, Lincoln's Inn Fields, London WC2A 3PX, U. K.

steroid-binding protein. Heyns and De Moor (1977) have called it prostatic-binding protein, in spite of the fact that to date, the protein has been shown to bind only steroids. The main advantages of studying the androgenic control of prostatic-binding protein are that it is secreted in vast amounts, accounting for approximately 40% of the total protein synthetic capacity of the gland and that its expression is extremely responsive to testosterone (Heyns *et al.*, 1977; Parker and Mainwaring, 1977; Parker *et al.*, 1978). Other steroid hormones and prolactin appear to play no role in its expression.

In fact, Fang and Liao (1971) first described the protein as a complex that bound dihydrotestosterone and was distinct from the androgen receptor because it was not translocated into the cell nucleus. It was then later rediscovered and renamed by many laboratories (Heyns and De Moor, 1977; Forsgren *et al.*, 1979; Parker *et al.*, 1978; Lea *et al.*, 1979; Chen *et al.*, 1979) and shown to bind a variety of steroids, albeit with low affinities. Whether this property of steroid binding is related to its function is unknown.

The protein consists of two subunits, each containing two polypeptides, one common and one unique to each subunit. Thus a polypeptide of mol. wt. 12,000 is linked by disulphide bridge(s) to a polypeptide of mol. wt. 10,000 in one subunit and 8,000 in the other subunit (Heyns *et al.*, 1978). These three polypeptides are synthesized as precursors, all of them containing a so-called "signal peptide" (Blobel and Dobberstein, 1975) which is subsequently cleaved when it is secreted and, in addition, the polypeptide of mol. wt. 12,000 is glycosylated with the result that it is actually larger than the primary translation product (Parker and Scrace, 1979; Peeters *et al.*, 1980). Thus when prostatic mRNA is translated in a cell-free system derived from wheat germ, three translation products of mol. wt. 9,000, 10,000, and 11,000 were produced. These were shown to be precursors of prostatic-binding protein by using microsomal membranes from dog pancreas which mimic the situation *in vivo* (Fig. 1). In view of the complexity of the biosynthesis of prostatic-binding protein, the polypeptides, mRNAs, and genes will be referred to by the mol. wt. of their primary translation products, namely, 9K, 10K, and 11K.

Basic questions concerning the expression and organization of the prostatic-binding-protein genes require that we clone cDNA molecules specific for the 9K, 10K, and 11K mRNAs to use as DNA probes for RNA analysis and for isolating DNA clones that contain the genomic sequences themselves. This chapter will, therefore, (1) outline an alternative method (Rougeon *et al.*, 1975) that we have used for cloning prostate cDNAs which offers two advantages over the usual method involving hairpin-loop formation followed by S1 nuclease digestion (Maniatis *et al.*, 1976; Williams, 1981); (2) discuss the steps in gene expression which may be influenced by androgens; and (3) briefly characterize prostate genomic DNA clones that contain the 9K, 10K, and 11K genes that will be essential if we are to investigate what factors are necessary for the expression of these genes in the ventral prostate.

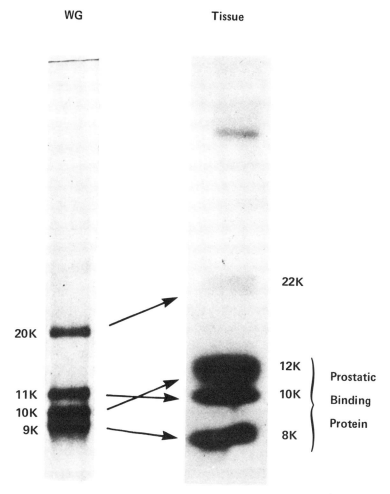

Figure 1. Expression of prostatic-binding protein. Autoradiography of [^{35}S]methionine-labeled proteins synthesized in tissue fragments and translated in a wheat germ cell-free system (WG) reveal four products. Three (mol. wt. 12,000, 10,000, and 8,000) make up prostatic binding protein and are translated as precursors of mol. wt. 10,000, 11,000, and 9,000, respectively, as indicated by arrows.

2. Cloning of Rat Ventral Prostate cDNAs

The most commonly used method for constructing double-stranded cDNA relies upon the ability of single-stranded cDNA to form a hairpin at its 3' end to prime the synthesis of a second DNA strand (Maniatis *et al.*, 1976; Williams, 1981). Although this method has been used successfully by many laboratories including our own, it results necessarily in the loss of DNA sequences corresponding to the 5' end of the mRNA when the hairpin loop is cleaved with S1 nuclease. An alternative strategy has been described (Rougeon *et al.,* 1975; Cooke *et al.,* 1980; Land *et al.,* 1981) in which it is possible to obtain more complete copies of the mRNA. A schematic diagram of the construction of double-stranded DNA is shown in Figure 2. The most important difference utilized in this method is the first step in which single-stranded cDNA is tailed with dCMP so that oligo dG can be used to prime the synthesis of the second DNA strand and DNA sequences corresponding to the hairpin loop are not lost. Another disadvantage of relying upon hairpin-loop formation is that it can lead to the artifactual generation of sequences in the cloned cDNA which are not present in the mRNA sequence (Sippel *et al.,* 1978; Richards *et al.,* 1979).

Using this strategy, double-stranded cDNA was annealed with a bacterial plasmid pAT153 (Twigg and Sherratt, 1980) that had been cut with Pst1 and tailed with oligo dG so that the Pst1 site was reformed. Such recombinant plasmids retain the tetracycline resistance loci but not ampicillin resistance and therefore, after transformation of *E. coli,* only bacteria capable of growth in the presence of tetracycline were selected. Preliminary screening of bacterial colonies was carried out by *in situ* hybridization (Grunstein and Hogness, 1975); in the case of the three prostate cDNAs (9K, 10K, and 11K) total ^{32}P-labeled cDNA was used as the probe. This is because these three mRNAs are the most abundant poly-(A)-containing RNAs in the ventral prostate and give rise to the predominant radiolabeled cDNA molecules which then result in the strongest hybridization

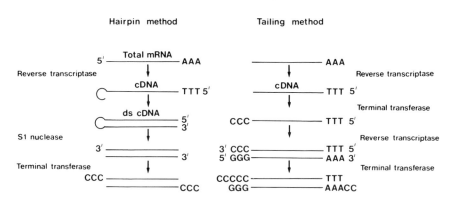

Figure 2.   Schematic diagram of the construction of double-stranded prostate cDNA. Two methods have been used to synthesize double-stranded DNA: one reaction relies on the ability of single-stranded cDNA to form a hairpin and act as a primer for the synthesis of the second strand; the other reaction involves the tailing of cDNA with oligo dC and the use of oligo dG as a primer for the synthesis of the second strand.

signals on the filters. Individual clones can be identified either by hybrid-arrested translation (Paterson *et al.*, 1977; Parker *et al.*, 1980) or by mRNA purification using recombinant plasmids bound to diazobenzyloxymethyl paper (Alwine *et al.*, 1977; Parker *et al.*, 1980). In this latter method, prostate mRNA is hybridized with immobilized DNA plasmids and, after washing to remove unhybridized RNA, bound RNA is eluted and translated in a wheat germ cell-free system. In addition, the size of the DNA inserts was estimated by digestion of the recombinant plasmids with PstI, since there are recognition sites for this enzyme at both ends of the insert. Several clones were almost the same size as that estimated for the mRNAs including a poly-(A)-tail (Parker *et al.*, 1980; Peeters *et al.*, 1980; Mansson *et al.*, 1980) and therefore it was likely that clones pA13, pA34, and pB44 represent complete copies of the 9K, 10K, and 11K mRNAs, respectively (Table I).

The DNA inserts in the recombinant plasmids were mapped with restriction enzymes and sequenced by the technique of Maxam and Gilbert (1977). In the case of pA34, the DNA insert consists of two Pst fragments, the larger of which contains most of the coding sequence (Fig. 3); the DNA sequence of this fragment agrees completely with the 10K protein sequence obtained by Peeters *et al.* (1981). Similarly, the DNA sequence for the 9K insert of pA13 agrees with the amino acid sequence obtained for the 9K protein (Peeters *et al.*, 1982). This represents excellent evidence that these cDNA clones are accurate copies of the prostatic-binding protein mRNAs which can be used as highly specific DNA probes for studies of both gene expression and organization.

The most notable feature, however, is the marked similarity between pA13 and pB44, with respect to the organization of their coding and noncoding regions and, more interestingly, their DNA sequence. Thus, extensive DNA sequence homologies occur both in the 5′ and 3′ noncoding regions and in the coding regions which result in a 50% amino acid sequence homology between the primary translation products and increase to 62% when semiconservative amino acids are included (Fig. 3A). The similarity in the DNA sequences in pA13 and pB44 suggests that the 9K and 11K genes have probably arisen by duplication of an ancestral gene followed by divergent evolution and may be of functional significance in view of the structure of prostatic steroid-binding protein (Parker *et al.*, 1982).

Table I. Relationships of Prostatic Proteins, mRNAs and Genes

Primary translation product	Processed polypeptide	mRNA size (nt)	cDNA plasmid	cDNA insert size (nt)	Gene
9,000	8,000	500–550	pA13	500	9K
10,000	12,000	650–700	pA34	610	10K
11,000	10,000	550–600	pB44	520	11K

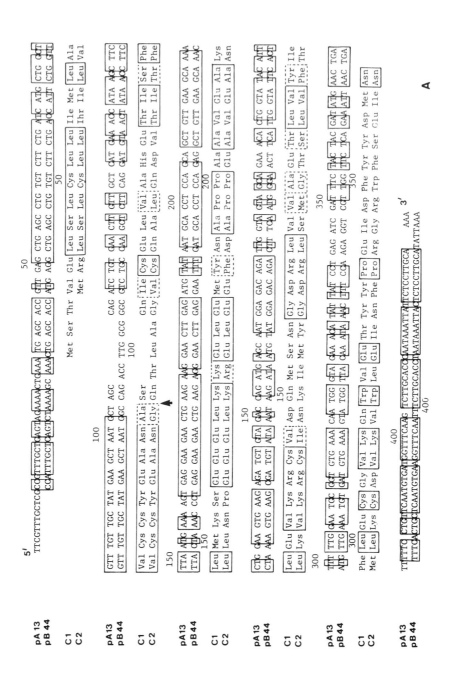

```
GAGTTCCTGATTTCTGTCTTGGACAACAGAACAACCCACAGGGACTGCCTCAAC

ATG AAG CTG GTG TTT CTA TTC TTG GTC ACC ATC CCC ATT TGC TGC TAT GCC
MET Lys Leu Val Phe Leu Leu Val Tnr Ile Pro Ile Cys Cys Tyr Ala

AGT GGT TCT GGC TGC AGT ATT CTA GAT GAG GTT ATT AGA GGT ACG ATT AAC TCG
Ser Gly Ser Gly Cys Ser Ile Leu Asp Glu Val Ile Arg Gly Thr Ile Asn Ser

ACT GTG ACG CTA CAT GAC TAT ATG AAA TTA GTT AAG CCA TAT GTA CAA GAT CAT
Tnr Val Tnr Leu His Asp Tyr MET Lys Leu Val Lys Pro Tyr Val Gln Asp His

TTT ACT GAA AAG GCT GTG AAG CAA TTC AAG CAG TGT TTT CTA GAT CAG ACC GAC
Phe Tnr Glu Lys Ala Val Lys Gln Phe Lys Gln Cys Phe Leu Asp Gln Tnr Asp

AAG ACT CTG GAA AAT GTT GGC GTG ATG GAG GCA ATA TTT AAC AGT GAA AGC
Lys Thr Leu Glu Asn Val Gly Val MET MET Glu Ala Ile Pne Asn Ser Glu Ser

TGT CAA CAG CCA TCC TAA ACA TCT ACA AGA TCT TTG GCC ACA GGA CTC CAG GAA
Cys Gln Gln Pro Ser

ACT GGC AAT GGC CAA GCA ACT GAT AAC ACA GAT CAT AAC TCT TCT TTC TTG AAC

CCC TTT TTC TAC CTA TAA AGT GCA AGA CGA TTG TTG AAA TCT CAA ATT TAT GTC

TTT CCA TTT TAT T
```

B

Figure 3. The DNA sequences and predicted amino acid sequences of pA13, pB44, and pA34. The DNA inserts were sequenced by the method of Maxam and Gilbert (1977) and the predicted amino acid sequence is included for (A) pA13 and pB44; (B) pA34. Gaps have been left to maximize homology; the solid boxes show perfect homologies and the dotted boxes show semiconservative amino acid homologies. The putative cleavage point of the signal peptide is represented by ↑.

3. Expression of RNA in Ventral Prostate

It has been established for several years that androgens stimulate gene expression primarily by altering mRNA levels. However, it had never been possible to isolate cDNA probes that are specific for individual prostatic-binding protein mRNAs because they are similar in size (Parker *et al.*, 1980; Peeters *et al.*, 1980). This makes their separation and the subsequent synthesis of specific cDNA molecules impossible. Cloning, of course, circumvents this problem and allows individual cDNA species to be isolated; these have been used to extend our earlier analysis of prostatic RNA. Very briefly, these studies demonstrate that the action of androgen in the ventral prostate is analogous to that of other steroids in their target tissues, namely 9K, 10K, and 11K mRNAs decline by about three orders of magnitude after hormone withdrawal and are restored rapidly by administration of testosterone (Parker *et al.*, 1980).

Alterations in cellular mRNA concentrations could result from hormonal effects on mRNA turnover or, alternatively, on RNA synthesis in the cell nucleus. Therefore, we quantitated the amounts of 9K, 10K, and 11K RNA in samples of nuclear RNA by "Northern blotting." In this technique RNA is separated by electrophoresis on agarose gels, transferred to diazobenzyloxymethyl paper, and quantitated *"in situ"* by using radiolabeled cDNA plasmids. Thus we were able to show that normal animals contained large amounts of 9K, 10K, and 11K RNA but these were absent in nuclear RNA from castrated animals and restored by testosterone treatment. Thus this further supports the notion that androgens stimulate primarily mRNA production. In addition, in the case of 10K RNA (Fig. 4), we observed two additional hybridizing bands of 1,500 and 4,000 nucleotides, which were also androgen-responsive, which we presume represent mRNA precursors. This is consistent with the fact that the 10K gene is split; analysis of the gene will be presented later.

Hormonal effects on transcription were analyzed by measuring RNA transcripts in prostate nuclei from normal and hormone-manipulated animals. Nuclei were incubated with [^{32}P]-UTP so that endogenous RNA polymerases were able to synthesize ^{32}P-labeled RNA that had been initiated *in vivo*. Specific transcripts were quantitated by hybridization to immobilized cDNA plasmids as described by McKnight and Palmiter (1979). The results that we obtained were similar for all three mRNAs and data for 10K mRNA are presented in Table II. 10K mRNA synthesis accounted for about 120 ppm in normal animals and 40–50 ppm in withdrawn animals, suggesting transcription rates are three times higher in the normal animal than in the hormone-withdrawn rat. This difference is not sufficient to account for the difference in the steady-state levels of mRNA and suggests that a major effect of testosterone is the stabilization of prostatic-binding protein RNAs. Thus testosterone appears to act on the rate of both transcription and turnover. In contrast, glucocorticoids particularly (Young *et al.*, 1977; Ringold *et al.*, 1977), and estradiol and progesterone in the chick oviduct (McKnight

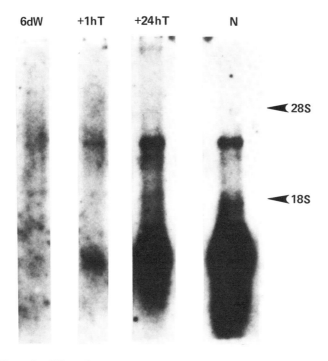

Figure 4.   Effect of androgens on nuclear 10K RNA. Nuclear RNA was isolated from animals withdrawn from androgen for 6 days, withdrawn animals treated with testosterone for 1 hr and 24 hr, and normal animals. After electrophoresis and transfer of the RNA to diazobenzyloxymethyl paper, the "northern" blot was hybridized with [^{32}P]-pA34 and 10K RNA was revealed by autoradiography.

Table II.   Transcription of 10K mRNA in Prostate Nuclei

Hormone status	Incorporation of [^{32}P]-UTP (p mol/μg DNA)	Input of [^{32}P]-RNA (cpm × 10^{-6})	[^{32}P]-RNA hybridized (cpm)	[^{3}H]-cRNA hybridized (%)	Rate of mRNA synthesis (ppm)
Withdrawn	0.033	2.81	47	35.6	47
Normal	0.050	2.00	122	52.1	117

Nuclei were incubated with [^{32}P]-UTP and the [^{32}P]-RNA transcripts were hybridized to filters containing immobilized pA34 and pAT. The rate of mRNA synthesis was calculated by subtracting nonspecific hybridization to pAT from the counts bound to pA34 and correcting for the efficiency of hybridization of the standard [^{3}H]-cRNA.

and Palmiter, 1979; Swaneck *et al.*, 1979) predominantly stimulate rates of transcription. Nevertheless it should be noted that, at least, estradiol and progesterone also affect mRNA turnover (Palmiter and Carey, 1974; McKnight and Palmiter, 1979). In fact, the action of testosterone most resembles the effect of the polypeptide hormone prolactin on casein gene expression in the mammary gland where mRNA transcription is stimulated two- to fourfold but the half-life of casein mRNA is increased 17- to 25-fold (Guyette *et al.*, 1979). Studies of turnover are notoriously difficult because few tissues or cells respond adequately to hormone *in vitro* for sufficient periods of time to carry out pulse–chase labeling experiments which are essential to investigate this problem in further detail. On the other hand, effects of hormones on transcription suggest that they interact directly with the gene and, therefore, several laboratories have isolated genomic DNA clones to investigate this possibility. An additional advantage of such clones is that we can use them to analyze their expression *per se* so that we can investigate the mechanism underlying tissue-specific gene expression.

4. Isolation of Prostatic-Binding Protein Genomic Clones

We have to elucidate the structure and organization of the prostatic-binding protein genes if we are to account for their expression in the ventral prostate but not other tissues, including target tissues for androgens. As a first step towards this goal we can isolate DNA clones from a so-called DNA library and analyze the organization of the genes by restriction enzyme mapping and electron microscopy. These libraries consist of fragments of DNA in the "arms" of lambda phage to produce recombinant phage that can be amplified in *E.coli*. Two such libraries, one utilizing EcoRI-generated DNA fragments and one using Hae III-generated DNA fragments, have kindly been provided by Drs. James Bonner and Tom Sargent and their colleagues (Sargent *et al.*, 1979).

From the size of the rat genome it can be calculated that about 200,000 recombinant phage would be required to cover the complete genome if the average size of the cloned fragments was 15 kb. These lambda clones were plated out to just subconfluence (10,000 per 14-cm petri dish) and replica-plated onto nitrocellulose filters. They were screened (Benton and Davis, 1977) first with ^{32}P-labeled total rat prostate cDNA because the three androgen-responsive mRNAs account for the largest proportion of the cDNA. All positive clones were identified (Fig. 5) and replated to purify them and the process repeated, being screened finally with specific cDNA clones so that individual genomic clones were identified. Thus far we have screened 600,000 EcoRI plaques and 600,000 Hae III plaques and identified two clones for each of 9K and 11K genes and 16 clones for the 10K gene (Table III).

Prior to detailed analysis of the clones, we established, using digestion with restriction enzymes, the DNA insert size and the degree of overlap between the

Probe 1:- ^{32}P-cDNA to total poly(A)-containing RNA

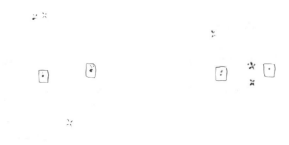

Probe 2:- nick-translated ^{32}P-labelled recombinant plasmid

Figure 5. Identification of recombinant phage that contain specific genomic DNA fragments. Recombinant phage were transferred to nitrocellulose paper and hybridized *in situ* (Benton and Davis, 1977) first with total [^{32}P]-cDNA and later, as the phage was purified, with specific [^{32}P]-cDNA plasmids. The autoradiographs from duplicate plates show positive clones.

Table III. Summary of Genomic Clones

Gene	Library	No. isolated	Insert size (Kb)
9K	Eco RI	1	19
	Hae III	1	13
10K	Eco RI	13	13–15 (3)
	Hae III	3	12–14 (10)
11K	Eco RI	2	15

Purified recombinant phage that had been identified with 9K, 10K, and 11K cDNA plasmids were digested with restriction enzyme, and the DNA insert size was determined.

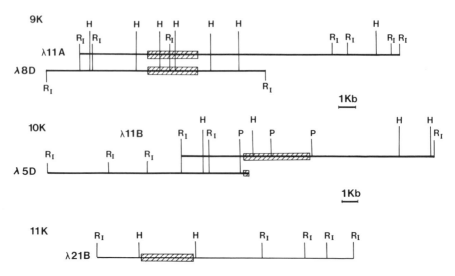

Figure 6. Restriction enzyme maps of prostatic-binding protein genes. The DNA inserts in each recombinant lambda phage were mapped with restriction enzymes (H, Hind III; P, Pst1; R, EcoRI) and overlapping clones indicated by the horizontal lines, were aligned. The coding regions (shaded boxes) were identified by Southern blotting and hybridization with [^{32}P]-cDNA plasmids.

various clones for each gene. The two 9K DNA clones were shown to overlap considerably (Fig. 6). Moreover, by hybridizing blots (Southern, 1975) with [^{32}P]-pA13 (9K cDNA) we were able to show that both clones contained similar coding sequences and therefore one, λ11A, was selected for further studies. The 10K clones were also shown to exhibit extensive overlap. Of the clones derived from the EcoRI DNA library, three were shown to contain a single 13.6-kb EcoRI DNA fragment while the remaining 10 contained this fragment plus an additional 1.6 Kb fragment. The Hae III clones also overlapped with this region and together with the EcoRI clones contained approximately 22 kb of the rat genome. The coding region was localized on the clones by hybridizing Southern blots with [^{32}P]-pA34 (10K cDNA) and, since all of them were similar, one λ11B, was selected for further studies. Finally, the two 11K clones were found to be similar and λ21B has been investigated in more detail. Thus, we appeared to have isolated DNA clones specific for all three prostatic-binding protein genes.

5. Characterization of Prostatic-Binding Protein Genes

It is extremely important to establish that the DNA in the recombinant clones truly represents the DNA in our Sprague–Dawley rats before detailed analysis of the genes is carried out. There is a great deal of manipulation of rat DNA to generate the DNA libraries that could lead to its artifactual modification. Potential artifacts include the ligation of noncontiguous restriction enzyme fragments when they were being ligated into the arms of lambda, and genetic rearrangements during propagation of the recombinant bacteriophage in *E.coli*. Such rearrangements would result in the existence of different restriction enzyme sites in the recombinant DNA clones than in rat DNA. Therefore, we analyzed blots of cloned DNA and cell DNA that had been digested with EcoRI, HindIII, Bam H1, and XbaI; results with the former enzyme only are presented in Figure 7.

Thus rat DNA contains two EcoRI fragments of 9.5 kb and 4.3 kb that contain 9K gene sequences; these two fragments were identified in λ11A suggesting that this clone contained the natural gene. Likewise, blots of the 11K gene were similar: both λ21B and rat DNA contained a single 9.8-kb EcoRI DNA fragment. Thus DNA clones λ11A and λ21B contain the natural 9K and 11K genes without any apparent modification.

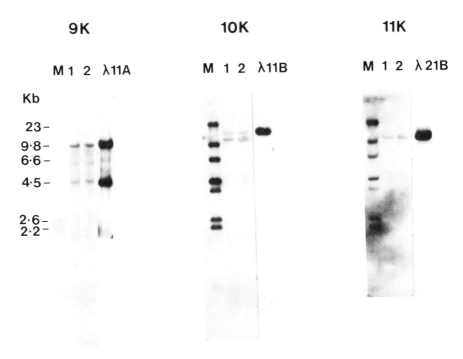

Figure 7.   Comparison of prostatic-binding protein genes in rat DNA and recombinant phage DNA. Southern blots of DNA from two individual rats (1 and 2) and from recombinant phage λ11A, λ11B, and λ21B were hybridized with [³²P]-cDNA plasmids specific for the 9K, 10K, and 11K genes, respectively. A Hind III digest of λ was used as DNA markers (M).

The situation for the 10K genes was less clear. Rat DNA contained two hybridizing EcoRI DNA fragments, only one of which was present in our clone. There are several explanations for this discrepancy: (1) our clone may contain only a portion of the 10K gene; (2) the rat tested may have been heterozygous for the 10K gene, mutations in the flanking sequence giving rise to allelic differences in the gene; or (3) there may be two copies of the 10K gene per haploid genome. We think from mapping studies and electron microscopy that the 10K gene consists of three exons separated by two intervening sequences so that the gene is approximately 4 kb. Since our genomic clones contain 22 kb rat DNA around this gene we think that the first possibility is unlikely because the existence of an additional exon in the gene would require the presence of an intervening sequence of at least 9 kb. Moreover we have evidence that there are distinct active genes in these flanking sequences adjacent to the 10K gene. To distinguish the second and third possibilities we have analyzed DNA from a large number of individual rats. All the rats contained both EcoRI fragments which suggests that they contain two copies of 10K per haploid genome. Polymorphism is unlikely because we would have expected that some of the animals would have been homozygous and contain only a single EcoRI DNA fragment. We are then left with the puzzle of why we failed to identify a recombinant phage that contained the additional gene, bearing in mind that we screened 1.2 $\times$ 10^6 phage plaques. It could have been lost when the DNA libraries were amplified or perhaps Sprague–Dawley rats at Cal–Tech do not contain two genes.

In any event, the existence of two genes in our colony of Sprague–Dawley rats raises a number of additional questions. First, are the two genes identical? because, thus far, we have only demonstrated that restriction enzyme sites in the flanking sequences are different. Second, are both genes expressed *in vivo* and, if not, does the clone we have isolated contain the active gene? We are endeavoring to answer these questions by mapping mRNA transcripts and by sequencing the genomic clones because we know the sequence of the 10K mRNA in the form of DNA in pA34. The protein sequence data suggest that only one gene is expressed but we cannot rule out the possibility that one gene is expressed at a higher level than the other and accounts for the bulk of the protein that has been sequenced.

6. Future Prospects

Recombinant DNA technology has enabled us to delineate which steps in gene expression are influenced by steroids and a great deal about the organization of hormonally responsive genes. Experiments to investigate the direct interaction of the hormone with the gene are in progress in several laboratories and, thus far, most success has come from studies of glucocorticoid-responsive genes using two completely different approaches. First, activated glucocorticoid receptors have been shown to bind selectively *in vitro* to a cloned DNA fragment from mouse mammary tumor virus whose transcription *in vivo* is regulated by glucocorticoids (Payvar *et al.*, 1981). A second approach being used to study gene expression is the introduction of genes into heterologous cells either by DNA-mediated transfection as pioneered by Axel and his coworkers (Pellicer *et al.*, 1980) or by direct microinjection into cells (Capechi, 1980) or fertilized oocytes (Wagner *et al.*, 1981). Thus, by cotransformation with thymidine kinase (TK) it has been possible to introduce glucocorticoid-responsive genes into TK⁻ L cells and stimulate their expression by the addition of glucocorticoid (Hynes *et al.*, 1981; Kurtz, 1981; Lee *et al.*, 1981). It is hoped that the glucocorticoid-responsive region of the gene will be identified if we delete portions of the DNA and test whether the gene retains hormone responsiveness. Unfortunately L cells do not have high titers of other steroid receptors and cell lines that contain androgen receptors will be required for the study of prostatic-binding protein genes. For example, S115 cells, which are derived from a spontaneous mouse mammary tumor, maintain androgen responsiveness in culture, but thus far, we have failed to isolate TK⁻ cells. An alternative strategy is to use a dominant selectable marker which confers a survival advantage upon the transformants. One such possibility is the use of a vector that contains the bacterial gene xanthine–guanine phosphoribosyl-transferase which enables mammalian cells to grow on xanthine (Mulligan and Berg, 1980). Finally, it may be possible to ligate putative regulatory regions of genes and coding regions of a second gene whose activity can be assayed *in vitro* such as interferon or whose function can be selected for such as thymidine kinase and dihydrofolate reductase (Lee *et al.*, 1981). If such fused genes are regulated by the hormone, it will be possible to identify the specific region of DNA that confers hormone-responsiveness.

REFERENCES

Alwine, J. C., Kemp, D. J., and Stark, G. R., 1977, Method for detection of specific RNAs in agarose gels by transfer to diazobenzylmethyl-paper and hydridisation with DNA probes, *Proc. Natl. Acad. Sci. USA* **74**:5350.

Benton, W. D., and Davis, R. W., 1977, Screening λgt recombinant clones by hybridisation to single plaques *in situ*, *Science* **196**:180.

Blobel, E., and Dobberstein, B., 1975, Transfer of proteins across membranes, *J. Cell. Biol.* **67**:835.

Capechi, M. R., 1980, High efficiency transformation by direct microinjection of DNA in cultured mammalian cells, *Cell* **22**:479.

Chen, C., Hirpakha, R. A., and Liao, S., 1979, Prostate α-protein: Subunit structure, polyamine binding, and inhibition of nuclear chromatin binding of androgen-receptor complex, *J. Steroid Biochem.* **11**:401.

Cooke, N. E., Cort, D., Weiner, R. J., Baxter, J. W., and Martial, J. A., 1980, Structure of cloned DNA complementary to rat prolactin mRNA, *J. Biol. Chem.* **255**:6502.

Fang, S., and Liao, S., 1971, Androgen receptors, *J. Biol. Chem.* **246**:16.

Forsgren, B., Bjork, P., Carlstrom, K., Gustafsson, J. A., Pousette, A., and Högberg, B., 1979, Purification and distribution of a major protein in rat prostate that binds estramustine, a nitrogen mustard derivative of estradiol-17β, *Proc. Natl. Acad. Sci. USA* **76**:3149.

Gronemeyer, H., and Pongs, O., 1980, Localization of ecdysterone on polytene chromosomes of *Drosophila melanogaster*, *Proc. Natl. Acad. Sci. USA* **77**:2108.

Grunstein, M., and Hogness, D. S., 1975, Colony hybridization: A method for the isolation of cloned DNAs that contain a specific gene, *Proc. Natl. Acad. Sci. USA* **72**:3961.

Guyette, W. A., Matusik, R. J., and Rosen, J., 1979, Prolactin-mediated transcription and post-transcriptional control of casein gene expression, *Cell* **17**:1013.

Heyns, W., and De Moor, B., 1977, Prostatic binding protein, *Eur. J. Biochem.* **78**:221.

Heyns, W., Peeters, B., and Mous, J., 1977, Influence of androgens on the concentration of prostatic binding protein (PBP) and its mRNA in rat prostate, *Biochem. Biophys. Res. Commun.* **77**:1492.

Heyns, W., Peeters, B., Mous, J., Rombauts, W., and De Moor, P., 1978, Purification and characterisation of prostatic binding protein and its subunits, *Eur. J. Biochem.* **89**:181.

Higgins, S. J., and Gehring, U., 1978, Molecular mechanisms of steroid hormone action, *Adv. Cancer Res.* **28**:313.

Hynes, N. E., Kennedy, N., Rahmsdorf, U., and Groner, B., 1981, Hormone-responsive expression of an endogenous proviral gene of mouse mammary tumor virus after molecular cloning and gene transfer into cultured cells, *Proc. Natl. Acad. Sci. USA* **78**:2039.

Kurtz, D. T., 1981, Hormonal inducibility of rat α_{2u} globulin genes in transfected mouse cells, *Nature* **291**:629.

Land, H., Grez, M., Hauser, H., Lindenmaier, W., and Schutz, G., 1981, 5'-Terminal sequences of eucaryotic mRNA can be cloned with high efficiency, *Nucleic Acids Res.* **9**:2251.

Lea, O. A., Petrusz, P., and French, F. S., 1979, Prostatein: A major secretory protein of the rat ventral prostate, *J. Biol. Chem.* **254**:6196.

Lee, F., Mulligan, R., Berg, P., and Ringold, G., 1981, Glucocorticoids regulate expression of dihydrofolate reductase cDNA in mouse mammary tumour virus chimaeric plasmids, *Nature* **294**:228.

Maniatis, T., Kee, S. G., Efstratiatis, A., and Kafatos, F. C., 1976, Amplification and characterisation of a β-globin gene synthesised *in vitro*, *Cell* **8**:163.

Mansson, P.-E. Silverberg, A. B., Gipson, S. H., and Harris, S. E., 1980, Purification of major abundance class of poly(A⁺)-RNA from rat ventral prostate, *Mol. Cell. Endocrinol.* **19**:229.

Maxam, A. H., and Gilbert, W., 1977, A new method for sequencing DNA, *Proc. Natl. Acad. Sci. USA* **74**:560.

McKnight, G. S., and Palmiter, R. D., 1979, Transcription regulation of ovalbumin and conalbumin genes by steroid hormones in chick oviduct, *J. Biol. Chem.* **254**:9050.

Mulligan, R. C., and Berg, P., 1980, Expression of a bacterial gene in mammalian cells, *Science* **209**:1422.

Palmiter, R. D., and Carey, N. H., 1974, Rapid inactivation of ovalbumin messenger ribonucleic acid after acute withdrawal of estrogen, *Proc. Natl. Acad. Sci. USA* **70**:2357.

Parker, M. G., and Mainwaring, W. I. P., 1977, Effects of androgens on the complexity of poly(A) RNA from rat prostate, *Cell* **12**:401.

Parker, M. G., and Scrace, G. T., 1979, Regulation of protein synthesis in rat ventral prostate: Cell-free translation of mRNA, *Proc. Natl. Acad. Sci. USA* **76**:1580.

Parker, M. G., Scrace, G. T., and Mainwaring, W. I. P., 1978, Testosterone regulates the synthesis of major proteins in rat ventral prostate, *Biochem. J.* **170**:115.

Parker, M. G., White, R., and Williams, J. G., 1980, Cloning and characterisation of androgen-dependent mRNA from rat ventral prostate, *J. Biol. Chem.* **255**:6996.

Parker, M. G., Needham, M., and White, R., 1982, Prostatic steroid binding protein: Gene duplication and steroid binding. *Nature* **298**:92.

Paterson, B. M., Roberts, B. E., and Kuff, E. L., 1977, Structural gene identification and mapping by DNA-mRNA hybrid-arrested cell-free translation, *Proc. Natl. Acad. Sci. USA* **74**:4370.

Payvar, F., Wrange, O., Carlstedt-Duke, J., Okret, S., Gustafsson, J.-A., and Yamamoto, K. R., 1981, Purified glucocorticoid receptors bind selectively *in vitro* to a cloned DNA fragment whose transcription is regulated by glucocorticoids *in vivo*, *Proc. Natl. Acad. Sci. USA* **78**:6629.

Peeters, B. L., Mous, J. M., Rombauts, W. A., and Heyns, W. J., 1980, Androgen-induced messenger RNA in rat ventral prostate, *J. Biol. Chem.* **255**:7017.

Peeters, B., Rombauts, W., Mous, J., and Heyns, W., 1981, Structural studies on rat prostatic binding protein, *Eur. J. Biochem.* **115**:115.

Peeters, B., Heyns, W., Mous, J., and Rombauts, W., 1982, Structural studies on rat prostatic binding protein, *Eur. J. Biochem.* **123**:55.

Pellicer, A., Robins, D., Wold, B., Sweet, R., Jackson, J., Lowy, I., Roberts, J. M., Sim, G. K., Silverstein, S., and Axel, R., 1980, Altering genotype and phenotype of DNA-mediated gene transfer, *Science* **209**:1414.

Richards, R. J., Shine, J., Ullrich, A., Wells, J. R. E., and Goodman, H. M., 1979, Molecular cloning and sequence analysis of adult chicken β globin cDNA, *Nucleic Acids Res.* **7**:1137.

Ringold, G. M., Yamamoto, K. R., Bishop, J. M., and Varmus, H. E., 1977, Increased rate of synthesis of viral RNA. Glucocorticoid-stimulated accumulation of mouse mammary tumour virus RNA, *Proc. Natl. Acad. Sci. USA* **74**:2879.

Rougeon, F., Kourilsky, P., and Mach, B., 1975, Insertion of a rabbit β-globin gene sequence into an E. coli plasmid, *Nucleic Acids Res.* **2**:2365.

Sargent, T. D., Wu, J.-R., Sala-Trepat, J. M., Wallace, R. B., Reyes, A. A., and Bonner, J., 1979, The rat serum albumin gene: Analysis of cloned sequences, *Proc. Natl. Acad. Sci. USA* **76**:3256.

Sippel, A. E., Land, H. Lindemaier, W., Nguyen-Hu, M. C., Wurtz, T., Timmis, K. N., Gresecke, K., and Schutz, G., 1978, Cloning of chicken lysozyme structural gene sequences synthesised *in vitro*, *Nucleic Acids Res.* **5**:3275.

Southern, E. M., 1975, Detection of specific sequences among DNA fragments separated by gel electrophoresis, *J. Mol. Biol.* **98**:503.

Swaneck, G. E., Nordstrom, J. L., Kreuzaler, F., Tsai, M. J., and O'Malley, B. W., 1979, Effect of estrogen on gene expression in chicken oviduct: Evidence for transcriptional control of ovalbumin gene, *Proc. Natl. Acad. Sci. USA* **76**:1049.

Twigg, A. J., and Sherratt, D., 1980, Trans-complementable copy-number mutants of plasmid Col. E1., *Nature* **293**:216.

Wagner, E. F., Stewart, T. A., and Mintz, B., 1981, The human β-globin gene and a functional viral thymidine kinase gene in developing mice, *Proc. Natl. Acad. Sci. USA* **78**:5016.

Williams, J. G., 1981, The preparation and screening of a cDNA clone bank, in: *Genetic Engineering* (R. Williamson, ed.), Vol. 1, pp. 1–59, Academic Press, London.

Yamamoto, K. R., Stampfer, M., and Tomkins, G. M., 1974, Receptors from glucocorticoid-sensitive lymphoma cells and two classes of insensitive clones: physical and DNA-binding properties, *Proc. Natl. Acad. Sci. USA* **71**:3901.

Young, H. A., Shih, T. Y., Scolnick, E. M., and Parks, W. P., 1977, Steroid induction of mouse mammary tumour virus: Effect upon synthesis and degradation of viral RNA, *J. Virol.* **21**:139.

11

Effects of Ovarian Steroid Hormones on the Brain and Hypophysis
Receptor Modulation and Chromatin Binding

Junzo Kato and Tsuneko Onouchi

1. Hormonal Modulation of Estrogen and Progesterone Receptors in the Rat Brain

1.1. Introduction

Progesterone plays a role in the central nervous system as a facilitator and inhibitor of sexual behavior and gonadotropin release in rodents. Estrogen-induced gonadotropin surges in female rats can be facilitated or inhibited by progesterone in the cycling rats (Everett, 1948; Zeilmaker, 1966; Brown-Grant *et al.*, 1972; Brown-Grant and Naftolin, 1972), and also be mimicked in the estrogen-treated ovariectomized rats by administration of progesterone (Caligaris *et al.*, 1971; Freeman *et al.*, 1976; Goodman, 1978), and in estrogen-treated prepubertal female rats of 28 days of age (Attardi, 1981). The effects of progesterone on sexual behavior in estrogen-primed female rats are also well-demonstrated (Meyerson, 1972; Feder and Marrone, 1977).

Junzo Kato ● Department of Obstetrics and Gynecology, Yamanashi Medical University, Tamaho, Nakakoma-gun, Yamanashi Prefecture, 409-38 Japan   *Tsuneko Onouchi* ● Department of Obstetrics and Gynecology, Teikyo University School of Medicine, Kaga, Itabashi, Tokyo, 173 Japan.

The biochemical mechanism of hormonal modulation has been investigated in terms of neural and hypophysial estrogen and progesterone receptors. However, the present situation is still conflicting and controversial. The enhanced nuclear retention of [³H]estradiol by progesterone was reported for the rat hypothalamus and hypophysis (Reuter and Lisk, 1976; Lisk and Reuter, 1977). On the other hand, no apparent effect by progesterone has been reported in cytoplasmic estradiol receptors (DeBold *et al.*, 1976; Pavlik and Coulson, 1976) and nuclear uptake of [³H]estradiol (DeBold *et al.*, 1976; Marrone and Feder, 1977). Recently Attardi (1981) reported no reduction in cytoplasmic and nuclear estradiol receptors in the hypothalamus-preoptic area (HPOA) and hypophysis in estrogen-treated 28-day-old female rats, indicating that modulation by progesterone of the estrogen-induced luteinizing hormone (LH) surge does not seem to result from effects on neural and hypophysial estrogen receptors. It is interesting, moreover, that progesterone can suppress the level of progesterone receptors in rat pituitary culture cells (Haug, 1979) and in rat HPOA and hypophysis (Schwartz *et al.*, 1979; Moguilewsky and Raynaud, 1979).

In an attempt to investigate the modulation by ovarian steroid hormones of estrogen and progesterone receptors in the central target tissues, the effects of progesterone or/and estradiol on estrogen and progesterone receptors were examined in the cytosols and nuclei of the HPOA, anterior hypophysis, and cerebral cortex of ovariectomized adult rats.

1.2. Materials and Methods

1.2.1. Chemicals

Redistilled and purified solvents and reagent-grade chemicals were used in all experiments. [2,4,6,7-³H]estradiol-17β (SA91.8 Ci/mmole), [17α-methyl-³H]17,21-dimethyl-19-nor-4,9-pregnadiene-3,20-dione (³H-R5020, specific activity 87.1 or 87.0 Ci/mmole) was obtained from New England Nuclear, Boston, Mass., and purified by thin-layer chromatography. Steroids were supplied by Sigma Chemical, St. Louis, Mo.

Sucrose and glycerol were purchased from Merck, Japan, and sodium EDTA and thioglycerol from Sigma Chemical, St. Louis, Mo. Dextran T70 (Pharmacia Fine Chemicals, Uppsala, Sweden) and Norit A (American Norit, Jacksonville, Florida) were used.

1.2.2. Schedule of Hormone Treatments

Adult rats of Wistar strain were bilaterally ovariectomized. From the 14th day after the operation the rats were injected subcutaneously once daily for 7

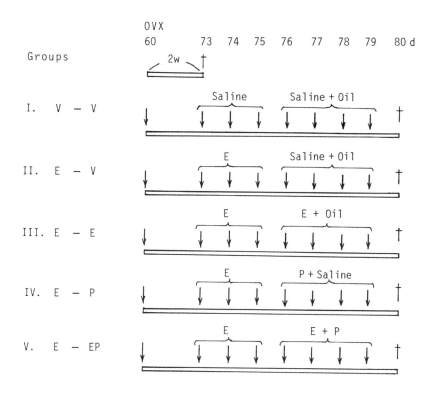

Figure 1. Experimental schedule for hormone treatment. V, Vehicle: saline or sesame oil; E, estradiol: 10 μg/ml saline; P, progesterone: 10 mg/0.4 ml sesame oil.

consecutive days with estradiol (10 μg/0.01 ml EtOH in 1 ml saline) and progesterone (10 mg/0.5 ml sesame oil) according to the regimen described in Fig. 1: group I, vehicle only; group II, estradiol and vehicle; group III, estradiol and estradiol; group IV, estradiol and progesterone; group V, estradiol, and estradiol plus progesterone. Each group of 8–11 rats was sacrificed by decapitation 24 hr after the last injection. Body weights of the rats at sacrifice were 258 ± 1.7 (SEM) gm (n = 177).

1.2.3. Preparation of Cytosols and Nuclei

The anterior pituitary, HPOA, and cerebral cortex, dissected as described previously (Kato and Villee, 1967) was dissected on an ice-cold glass plate. The tissues were rinsed in ice-cold 10 mM Tris-HCl buffer (pH 7.4) containing 1 mM sodium EDTA, 12 mM thioglycerol, and 10% glycerol (TETG) and were blotted on filter paper. They were weighed and suspended in the TET solution as follows: 3 vol (g/ml) for HPOA and cerebral cortex, 10 vol for pituitary. The tissues were homogenized with a Teflon pestle in a homogenizer (Ikemoto, Tokyo; 5 strokes, 600–1000 rpm, 3 min) at 4°C. The homogenates were centrifuged at 800g for 20 min to obtain a nuclear fraction and low-speed supernatant fraction. The nuclear fraction was immediately suspended in 0.25 M sucrose-TMKG buffer (0.25 M sucrose, 10 mM Tris-HCl, 12 mM thioglycerol, 2 mM MgCl$_2$, 10 mM KCl, and 10% glycerol, pH 7.5) and centrifuged at 800g for 20 min. The pellet was resuspended and sedimented three more times; once in 0.25 M sucrose–TMKG buffer containing 0.5% Triton X-100 and twice in 0.25 M sucrose–TMKG buffer. The final pellet was suspended in 0.25 M sucrose–TMKG buffer. The low-speed supernatant fraction was spun at 105,000g for 1 hr in a Hitachi 65P model ultracentrifuge (Hitachi Corp., Tokyo) to yield the resulting high-speed supernatant (cytosol). The cytosol fraction was then treated with dextran-coated charcoal to remove free hormone. All subsequent procedures, unless otherwise noted, were performed at 4°C.

1.2.4. Measurement of Cytosol Receptors

1.2.4.1. Available, Occupied, and Total Cytosol Estradiol Binding Sites. Aliquots of cytosol were incubated in duplicate with several concentrations of [^{3}H]estradiol (in the range from 0.12 to 8.57 nM), to determine available (unoccupied) binding sites, or with [^{3}H]estradiol plus a 100-fold excess of diethylstilbestrol (DES), to determine nonspecific binding, in a total volume of 310 µl. Tubes were kept at 0–4°C for 17–18 hr. Separation of bound and free [^{3}H]estradiol was performed, according to the modified method of Korenman and Dukes (1970) and McGuire (1975), with dextran–charcoal (1 ml, 0.0025% Dextran T-70, 0.25% Norit A in TETG, pH 8.0). One ml of the charcoal-treated supernatants was counted. Total binding sites were determined by incubation of cytosols at 25°C for 2 hr (Katzellenenbogen *et al.*, 1973).

Radioactivity was measured in an Aloka Model LSC-900 scintillation counter (Aloka Corp., Mitaka, Tokyo) in a Triton–toluene–Omnifluor (NEN, Boston, Mass.) system. The counting efficiency was 38%. Specific binding was determined by subtracting nonspecific binding from total binding. The number of binding sites (NBS) and dissociation constants (Kds) were determined by the Scatchard plot (Scatchard, 1949).

1.2.4.2. Progesterone Receptors. The cytosols (300 µl) were incubated at 4°C for 16 hr with various levels of concentration of [^{3}H]-R5020 (0.12–1.98 nM) with or without 100-fold excess of unlabeled R5020 (Kato and Onouchi, 1979). Dextran-coated charcoal solution (1 ml, 0.0025 g% Dextran T70, 0.25 g% Norit A in TETG, pH 8.0) was added and mixed well for 10 min in a vortex, followed by centrifugation at 1500 xg for 10 min. Aliquots of the supernatant were counted.

1.2.5. Measurement of Nuclear Estradiol and Progesterone Binding Sites

Measurement of nuclear estradiol binding sites was carried out by the modified method of Anderson *et al*. (1972). Nuclear pellets (50–100 μg, DNA) were dispersed in 0.3 ml (HPOA, pituitary) 0.25 M sucrose–TMKG buffer, and incubated in duplicate with [^{3}H]estradiol (8.0 or 8.6 nM) in the presence of a 100-fold excess of DES for 1 hr at 30°C. After incubation, the tubes were kept in icc for 30 min, and washed as previously described. The nuclear pellet was added with 1.5 ml of ethanol for extraction of bound nuclear binding sites. Aliquots of ethanol solution were counted in a toluene–Omnifluor system.

Nuclear progesterone receptors were determined by the low-temperature exchange assay (Kato and Onouchi, 1979). The dispersed nuclear pellet was incubated in duplicate with [^{3}H]-R5020 (7.6 or 7.7 nM) in the presence or absence of unlabeled R5020 for 2 hr at 10°C.

1.2.6. Miscellaneous Assays

Cytosol protein concentrations were measured by the method of Lowry *et al*. (1951) using crystallized bovine serum albumin as standard.

DNA was quantitated by the method of Burton (1956), using calf thymus DNA as the standard.

1.2.7. Statistical Analysis

To evaluate significant differences between groups, Student's *t* test (two-tailed) for estradiol and progesterone receptor levels and *K*ds for [^{3}H]estradiol and progesterone binding were determined.

1.3. Results

Hormonal modulation of estradiol and progesterone receptors in the central targets were examined in the cytosol and nuclei from the HPOA, anterior hypophysis, and cerebral cortex from ovariectomized rats treated with estrogen or/and progesterone according to the regimen described. (Fig. 1).

1.3.1. Effects on Estradiol Receptors (ERs)

Data on available, occupied, and total cytosol, nuclear, and total estradiol receptors in the HPOA and anterior hypophysis are summarized in Table I.

1.3.1.1. Hypothalamic ER Level. Effects of Estradiol.   Estradiol injections for 7 consecutive days increased the concentrations of nuclear ER (ERn) in the HPOA (V–V group vs. E–E group, P < 0.01; V–V group vs. E–EP group, P < 0.05), but did not affect statistically significantly the levels of available,

Table I. Effects of Estradiol and Progesterone on Estradiol Receptors in the
Hypothalamus–Preoptic Area and Anterior Hypophysis of Adult Ovariectomized Rats

Groups for hormone treatment	Cytosol ER(ERc) (fmol/mg prot.)			Nuclear ER (fmol/mg P)	Total ERs (fmol/mg P)
	Unoccupied	Occupied	Total ERc		
Hypothalamus					
I.V-V[a]	32.5 ± 3.66[b]	1.78 ± 1.38	33.9 ± 3.99	0.63 ± 0.37	34.6 ± 4.33
II.E-V	24.5 ± 5.35	4.13 ± 2.52	28.7 ± 4.66	0.88 ± 0.25	29.5 ± 4.42
III.E-E	29.1 ± 2.98	1.33 ± 1.13	29.3 ± 1.81	3.75 ± 0.62	33.1 ± 2.02
IV.E-P	18.5 ± 2.92	2.18 ± 1.46	20.0 ± 2.03	0.98 ± 0.41	21.0 ± 2.00
V.E-EP	22.7 ± 1.53	0.55 ± 0.49	21.5 ± 2.40	2.80 ± 0.58	24.0 ± 2.93
Hypophysis					
I.V-V	157.7 ± 8.17	0	149.4 ± 11.5	12.70 ± 1.19	162.0 ± 10.8
II.E-V	176.9 ± 20.4	0	155.5 ± 17.7	9.78 ± 1.04	165.2 ± 17.5
III.E-E	121.9 ± 14.3	9.15 ± 5.46	127.8 ± 15.3	22.80 ± 2.72	150.6 ± 13.8
IV.E-P	115.6 ± 19.1	5.95 ± 2.10	121.6 ± 17.0	9.33 ± 0.76	130.9 ± 15.9
V.E-EP	108.2 ± 7.97	10.50 ± 4.40	118.2 ± 11.1	17.30 ± 3.59	135.4 ± 8.7

[a] V; Vehicle, E; estradiol(10µg), P; progesterone(10 mg).
[b] Mean ± SEM of four determinations.

occupied, or total cytosol ERs. No significant effects of estrogen on total ERs
(cytosol and nuclear) were found.

Effects of progesterone. No statistically significant effects of P on unoccupied
and total cytosol ERs were found, except for comparison of V–V group to E–P
group ($P < 0.05$). There was also no difference in occupied cytosol ER values
between any combination of pair groups. Progesterone seems unlikely to have
decreasing effect on the hypothalamic cytosol ERs. The concentrations of nuclear
ER were not affected by progesterone. The values of total ERs were, however,
decreased by progesterone (E–E group vs. E–EP group, $P < 0.05$), indicating
that progesterone has the antiestrogenic action through its lowering effect on
total ER level in the hypothalamus.

1.3.1.2. Hypophysial ER Level. Estrogen increased nuclear ER level
(E–V group vs. E–E group, $P < 0.05$), but not the cytosol ER level (Table I).
There was no alteration in the concentrations of total ERs in the tissue. Although
nuclear ER level was lowered between E–E group and E–P group ($P < 0.05$),
no significant effect was found between the following pair groups (E–V group
vs. E–P group; E–V group vs. E–EP group).

Progesterone tended to lower the concentrations of unoccupied and total
cytosol ER, but the difference was not statistically significant. Occupied cytosol
ER levels were increased (V–V group vs. E–P group; E–V vs. E–P group). The
value of total ERs was, however, not affected by progesterone injections.

1.3.1.3. Cerebral Cortex. In the cerebral cortex the levels of cytosol ER and nuclear ER were very low. No significant changes were found by the hormonal administration (data not shown).

1.3.2. Effects on Progesterone Receptors (PRs)

Data on the hormonal effects on progesterone receptors in the hypothalamus and hypophysis are shown in Table II.

1.3.2.1. Hypothalamic PR Level. Effects of Estradiol. Estradiol markedly increased the concentrations of cytosol, nuclear, and total ERs (V–V group vs. E–E group; E–V group vs. E–E group). The present data are in good agreement with previous findings that estrogen can induce progesterone receptors in a dose-dependent fashion in the hypothalamus and hypophysis of the rat (Kato and Onouchi, 1977; Kato *et al.*, 1978; MacLusky and McEwen, 1978; Moguilewsky and Raynaud, 1979). It is noteworthy that nuclear PR was increased with estradiol administration. Estrogen-induced cytosol PRs in the hypothalamus might be translocated by endogenous progesterone from the adrenals during estrogen administration.

Table II. Effects of Estrogen and Progesterone on
Progesterone Receptors in the Hypothalamus–Preoptic
Area, Anterior Hypophysis, and Cerebral Cortex of Adult
Ovariectomized Rats

Groups for hormone treatment	Cytosol PR (fmol/mg P)	Nuclear PR (fmol/mg P)	Total PR (fmol/mg P)
Hypothalamus			
I.V-V	10.5 ± 2.11^a	0.83 ± 0.30	11.7 ± 2.38
II.E-V	12.7 ± 2.09	0.63 ± 0.69	13.5 ± 2.02
III.E-E	25.4 ± 1.17	2.08 ± 0.28	27.5 ± 1.41
IV.E-P	9.2 ± 2.57	3.75 ± 0.93	12.9 ± 3.27
V.E-EP	16.2 ± 2.57	6.05 ± 1.28	22.2 ± 3.05
Anterior Hypophysis			
I.V-V	9.5 ± 4.77	0.97 ± 0.62	10.5 ± 5.08
II.E-V	13.2 ± 1.13	2.25 ± 0.98	15.5 ± 0.91
III.E-E	85.8 ± 10.3	4.88 ± 0.77	90.7 ± 9.53
IV.E-P	11.3 ± 0.41	6.33 ± 2.06	14.8 ± 1.92
V.E-EP	28.4 ± 2.93	16.6 ± 2.14	45.0 ± 4.93
Cerebral Cortex			
I.V-V	15.2 ± 2.63	0.55 ± 0.55	15.8 ± 3.03
II.E-V	15.9 ± 1.73	1.23 ± 0.58	17.1 ± 2.26
III.E-E	17.5 ± 2.12	0.80 ± 0.38	18.3 ± 2.39
IV.E-P	12.4 ± 1.64	2.85 ± 1.19	15.3 ± 2.67
V.E-EP	13.7 ± 2.77	2.48 ± 0.88	16.1 ± 1.87

[a] Mean ± SEM of four determinations.

Effects of progesterone. When progesterone alone was injected following the estrogen priming, progesterone intended to decrease cytosol PR and increase nuclear PR (E–V group vs. E–P group); the differences were not statistically significant. Progesterone alone did not affect the values of total PR in the hypothalamus. When progesterone was injected in combination with estradiol, progesterone caused reduction of cytosol PR and increase of nuclear PR (E–E group vs. E–EP group), but no significant effect was found in the concentrations of total PRs.

    1.3.2.2. Hypophysial Level. Estradiol showed a marked induction of hypophysial cytosol, nuclear, and total PRs. Elevated levels of nuclear PR were, as in the hypothalamus, observed in the anterior hypophysis (V–V group vs. E–E, or E–EP).

    Progesterone alone did not alter the concentrations of cytosol, nuclear, and total PRs. When progesterone was injected simultaneously with estradiol, progesterone resulted in marked reduction of cytosol PR (E–E group vs. E–EP, $P < 0.01$) with increase in nuclear PR (E–E group vs. E–EP group, $P < 0.01$). The values of total PRs were also depressed by progesterone ($P < 0.01$).

1.3.3. Cerebral Cortex

    There was no statistically significant difference in the concentrations of cerebral cortical PRs between any pair of five experimental groups.

1.4. Discussion

    Despite the well-established action of progesterone on facilitation and inhibition of sexual behavior and gonadotropin (Everett, 1948; Zeilmaker, 1966; Brown-Grant *et al.*, 1972; Brown-Grant and Naftolin, 1972; Caligaris *et al.*, 1971; Freeman *et al.*, 1976; Goodman, 1978), much of its biochemical mechanisms still remain unclear. While progesterone depresses the replenishment of the estrogen receptors and the subsequent nuclear translocation in rat uterus (Hsueh *et al.*, 1976; Pavlik and Coulson, 1976), its modulatory effects on estrogen and progesterone receptors in the central tissues have been conflicting and controversial (DeBold *et al.*, 1976; Pavlik and Coulson, 1976; Reuter and Lisk, 1976; Lisk and Reuter, 1977; Marrone and Feder, 1977; Attardi, 1981).

    As shown in Table I, progesterone, injected with estradiol, decreased total ERs in the HPOA. It is suggested that progesterone has an antiestrogenic action through its effect on lowering total ER level in the HPOA. The present data are inconsistent with the enhanced uptake of [³H]estradiol in the hypothalamus (Reuter and Lisk, 1976; Lisk and Reuter, 1977) and no apparent effects on cytosol and nuclear ERs in the HPOA (DeBold *et al.*, 1966; Pavlik and Coulson, 1976; Marrone and Feder, 1977; Attardi, 1981). Attardi (1981) reported no changes in cytosol and nuclear ERs in the HPOA of 28-day-old rats, indicating that modulation by progesterone of the estrogen-induced LH surge does not seem to result from effects on neural and hypophysial estrogen receptors. The discrepancy might be due to experimental conditions such as animal age, the presence or absence of castration, regimen and route of injection, and dosages of progesterone, etc. It is assumed from our data that inhibition by progesterone of the estrogen-induced LH surge seems to be due to depression of ERs in the HPOA. In the hypophysis, unlike the hypothalamus, there is no apparent effect of progesterone on ER. The modulatory effect of progesterone seems unlikely to be associated with the change in the level of ER in the hypophysis.

The concentrations of hypophysial total PRs were depressed by administration of progesterone (Table II). This is in good agreement with inhibitory effects of progesterone on progesterone receptors in culture cells *in vitro* (Haug, 1979) and *in vivo* (Schwartz *et al.*, 1979; Moguilewsky and Raynaud, 1979). Since the synthesis of progesterone receptors can be regulated at least in part by estrogen (Kato and Onouchi, 1977; Kato *et al.*, 1978; MacLusky and McEwen, 1978; Moguilewsky and Raynaud, 1979), there is the possibility that reduction in progesterone receptors is due to depression of hypophysial ERs by progesterone. In the present experiments, however, there is no effect of progesterone on hypophysial ERs. These suggest the existence of regulation of progesterone receptors by progesterone itself. Suppression of progesterone receptors in the preoptic–anterior hypothalamus (Schwartz *et al.*, 1979; Moguilewsky and Raynaud, 1979) is inconsistent with our negative data on the HPOA. The discrepancy might be partly explained by different dissection of the hypothalamus.

The present data on marked elevation of hypothalamic and hypophysial progesterone receptors by estrogen confirm and extend previous reports (Kato and Onouchi, 1977; Kato *et al.*, 1978; MacLusky and McEwen, 1978; Moguilewsky and Raynaud, 1979). It is noteworthy that nuclear PR is increased with estradiol administration. Estrogen-induced cytosol PRs in the hypothalamus might be translocated by endogenous progesterone from the adrenals during estrogen administration.

2. Chromatin Binding in the Brain and Hypophysis

2.1. Introduction

Estrogen-induced progesterone receptors in the hypothalamus and hypophysis are considered as evidence for the synthesis of specific protein which is an end product of estrogen-receptor-mediated action of the steroid hormone on the gene. Analysis of the gene action of estrogen is important in elucidation of the action of the hormone on the brain. Little is known, however, about even an initial biochemical process such as chromatin binding of estrogen–receptor complexes in the central target tissues.

In an attempt to learn whether the hypothalamic and hypophysial cytosol estrogen receptors can bind to the chromatin as well as how specificity of the target cells is determined on the level of chromatin, we examined the binding of hypothalamic and hypophysial cytosol estradiol receptors to chromatins obtained from estrogen-sensitive central tissues (the HPOA or anterior hypophysis), peripheral tissue (the uterus), and nonresponsive tissue (the spleen) from 27-day-old rats.

2.2. Methods and Materials

Cytosols (2 mg of protein) from the HPOA and anterior hypophysis were incubated with [³H]estradiol (17.3 nM, SA 109 Ci/mmole) for 1 hr at 0°C, followed by 15 min at 25°C. The charged cytosols, treated with dextran-coated charcoal solution for 15 min at 0°C, were further incubated with chromatin (35–57 μg) obtained from the HPOA, anterior hypophysis, uterus, or spleen for 1 hr at 0°C in a final volume of 0.5 ml containing 1.15 M NaCl. Chromatin from the brain, and from uterus and spleen was prepared according to Shaw and Huang (1970), and Spelsberg *et al.* (1971), respectively. The millipore filter assay was carried out according to the method of O'Malley *et al.* (1973).

Chromatin binding was calculated from subtraction of nonspecific binding (without the cytosols) from radioactivity in incubation of chromatin with the respective cytosols and expressed as dpm ([³H]estradiol binding)/μg DNA.

2.3. Results and Discussion

As shown in Table III, the hypothalamic cytosol–[³H]estradiol complexes bound the chromatins from all tissues tested. If the binding to the chromatin

Table III. Binding of [³H]Estradiol-Charged Hypothalamic or Hypophysial Cytosol to Chromatins from the Hypothalamus, Anterior Hypophysis, Uterus, and Spleen

Charged cytosol[a]	Binding to chromatins (dpm/μg DNA)			
	Hypothalamus[b]	Anterior hypophysis	Uterus	Spleen
Hypothalamus	31.1 ± 3.13[c]	25.2 ± 1.88	16.1 ± 1.92	18.5 ± 0.87
Anterior Hypophysis	229.8[d]	182.6	153.7	158.2

[a] The cytosols (2 mg of protein) were labeled *in vitro* with [³H]estradiol (17.3 nM, SA 100 Ci/mmole). Binding of the charged cytosols to chromatins was carried out. Detail of these procedures are described in text.
[b] Hypothalamus: hypothalamus–preoptic area.
[c] Mean and SEM of four determinations.
[d] Mean of two determinations.

from the hypothalamus is taken as 100% and compared with that to other tissue chromatins, the values for the hypophysis, uterus, and spleen were 81, 51, and 60%, respectively. No significant difference in the chromatin binding was seen between the hypothalamus and hypophysis, suggesting no tissue-specificity in these central target tissues. In contrast, the binding of the charged hypothalamic cytosols to the chromatin from the uterus or spleen was markedly decreased (P < 0.05).

The chromatin binding of charged hypophysial cytosols was essentially similar to that of the hypothalamic ones; the values for the hypothalamus, uterus, and spleen were 80, 67, and 69%, respectively.

These results suggest that differential chromatin binding of the hypothalamic and, probably, hypophysial cytosol estrogen–receptor complexes exists in the estrogen-sensitive central (hypothalamo-hypophysial) target, and the peripheral target uterus and nontarget tissues (spleen). It is noticeable that there is no tissue-specificity in chromatin binding between the hypothalamus and hypophysis. Since the lack of organ specificity of estrogen cytosol receptors has been reported in the hypothalamus, hypophysis, and uterus (Notides, 1970; Talley *et al.*, 1975; Kato, 1978), the tissue specificity of the target cells of the hypothalamus and hypophysis may be determined at least in part at the level of chromatin. It is further tempting to speculate that the hypothalamic and, probably, hypophysial chromatin binding sites, that is, "acceptor sites," can differentiate and bind their cytosol receptor binding sites.

3. Summary and Conclusion

3.1. Modulation of Ovarian Steroid Receptors

In order to investigate the modulation by ovarian steroid hormones of estrogen and progesterone receptors in the central target tissues, the effects of progesterone or/and estradiol on estradiol and progesterone receptors were examined in the cytosols and nuclei of the HPOA, anterior hypophysis, and cerebral cortex of ovariectomized adult rats.

Progesterone depressed the concentrations of total estradiol receptors in the HPOA, but not in the anterior hypophysis, suggesting its antiestrogenic effect probably occurs through lowering the level of total estrogen receptors in the HPOA. The values of total progesterone receptors in the anterior hypophysis were suppressed by progesterone, indicating the presence of down-regulation of progesterone receptors in the tissue. Estrogen-induced increase in progesterone receptors was confirmed in the HPOA and anterior hypophysis. No modulation of estradiol receptors by estrogen was observed. These results suggest that ovarian steroids can modulate their receptors in the HPOA and hypophysis.

3.2. Chromatin Binding Study

In attempting to learn how the specificity of estradiol binding is determined at the level of chromatin, we investigated the binding of the hypothalamic (HPOA) or hypophysial cytosol estradiol complexes to chromatins from the estrogen-sensitive central (HPOA and anterior hypophysis), peripheral (uterus), or nontarget tissue (spleen). The binding of hypothalamic and hypophysial cytosols to the respective chromatins was greater than that to chromatins from the uterus or spleen. It is suggested that differential chromatin binding of the hypothalamic and, probably, hypophysial cytosol receptor–estradiol complexes exists in the central and peripheral target cells. The tissue specificity of the central target cells may be determined at least in part at the level of chromatin.

ACKNOWLEDGMENTS. The authors express their appreciation to Dr. S. Okinaga and Dr. K. Arai for generous assistance, and Ms. Y. Onda for preparation of the manuscript. Supported by Grants-in-Aid No. 56120006, No. 57113007 from the Japan Ministry of Education, Science, and Culture (J.K.).

REFERENCES

Anderson, J., Clark, J. H., and Peck, Jr. E. J., 1972, Oestrogen and nuclear binding sites. Determination of specific sites by ^{3}H-oestradiol exchange, *Biochem. J.* **126**:561.
Attardi, B., 1981, Facilitation and inhibition of the estrogen induced luteinizing hormone surge in the part by progesterone: Effects on cytoplasmic and nuclear estrogen receptors in the hypothalamic preoptic area, pituitary, and uterus, *Endocrinology* **108**:1487.
Brown-Grant, K., and Naftolin, F., 1972, Facilitation of luteinizing hormone secretion in the female rat by progesterone. *J. Endocrinol.* **53**:37.
Brown-Grant, K., Corker, C. S., and Naftolin, F., 1972, Plasma and pituitary luteinizing hormone concentrations and peripheral plasma oestradiol concentration during early pregnancy and after the administration of progestational steroids in the rat, *J. Endocrinol.* **53**:31.
Burton, K., 1956, A study of the conditions and mechanisms of the diphenylamine reaction for the colorimetric estimation of deoxyribonucleic acid, *Biochem. J.* **62**:315.
Caligaris, L., Astrada, J. J., and Taleisnik, S., 1971, Biphasic effect of progesterone on the release of gonadotropin in rats, *Endocrinology* **89**:331.
DeBold, J. F., Martin, J. V., and Whalen, R. E., 1976, The excitation and inhibition of sexual receptivity in female hamsters by progesterone: Time and dose relationships, neural localization and mechanisms of action, *Endocrinology* **99**:1519.
Everett, J. W., 1948, Progesterone and estrogen in the experimental control of ovulation time and other features of the estrous cycle in the rat, *Endocrinology* **43**:389.
Feder, H. H., and Marrone, B. L., 1977, Progesterone: Its role in the central nervous system as a facilitator and inhibitor of sexual behavior and gonadotropin release, *Ann. N. Y. Acad. Sci.* **286**:331.
Freeman, M. C., Dupke, K. C., and Croteau, C. M., 1976, Extinction of the estrogen-induced daily signal for LH release in the rat: A role for the proestrous surge of progesterone, *Endocrinology* **99**:223.

Goodman, R. L., 1978, A quantitative analysis of the physiological role of estradiol and progesterone in the control of tonic and surge secretion of LH in the rat, *Endocrinology* **102**:142.

Haug, E., 1979, Progesterone suppression of estrogen-stimulated prolactin secretion and estrogen receptor levels in rat pituitary cells, *Endocrinology* **104**:429.

Hsueh, A. J. W., Peck, Jr. E. J., and Clark, J. H., 1976, Control of uterine estrogen receptor levels by progesterone, *Endocrinology* **98**:438.

Kato, J., 1978, Brain receptors for sex steroid hormones and their characterization and functional implications, in: *Brain–Endocrine Interaction III. Neural Hormones and Reproduction* (D. E. Scott, G. P. Kozlowski, F. Collins, and A. Weindl, eds.), pp. 287–301, Karger, Basel.

Kato, J., and Onouchi, T., 1977, Specific progesterone receptors in the hypothalamus and anterior hypophysis of the rat, *Endocrinology* **10**:920.

Kato, J., and Onouchi, T., 1979, Nuclear progesterone receptors and characterization of cytosol receptors in the rat hypothalamus and anterior hypophysis, *J. Steroid Biochem.* **11**:845.

Kato, J., and Villee, C. A., 1967, Preferential uptake of estradiol by the anterior hypothalamus of the rat, *Endocrinology* **80**:567.

Kato, J., Onouchi, T., and Okinaga, S., 1978, Hypothalamic and hypophysial progesterone receptors: Estrogen priming effect, localization and isolation of nuclear receptors, *J. Steroid Biochem.* **9**:415.

Katzenellenbogen, J. A., Johnson, Jr. H. J., and Carlson, K. E., 1973, Studies on the uterine cytoplasmic estrogen binding protein. Thermal stability and ligand dissociation rate. An assay of empty and filled sites by exchange, *Biochemistry* **12**:4092.

Korenman, S. G., and Dukes, B. A., 1970, Specific estrogen binding by the cytoplasm of human breast carcinoma, *J. Clin. Endocrinol. Metab.* **30**:639.

Lisk, R. D., and Reuter, L. A., 1977, *In vivo* progesterone treatment enhances [^{3}H]estradiol retention by neural tissue of the female rat, *Endocrinology* **100**:1652.

Lowry, O. H., Rosebrough, N. J., Farr, A. L., and Randall, R. J., 1951, Protein measurement with the Folin phenol reagent, *J. Biol. Chem.* **193**:265.

MacLusky, N. J., and McEwen, B. S., 1978, Oestrogen modulates progestin receptor concentrations in some rat brain regions but not in others, *Nature* **274**:276.

Marrone, B. L., and Feder, H. H., 1977, Characteristics of [^{3}H]estrogen and [^{3}H]progestin uptake and effects of progesterone on [^{3}H]estrogen uptake in brain anterior pituitary and peripheral tissues of male and female guinea pigs, *Biol. Reprod.* **17**:42.

McGuire, W. L., 1975, Quantitation of estrogen receptors in mammary carcinoma, in: *Methods in Enzymology*, vol. 36: *Hormone Action* (B. W. O'Malley and J. G. Hardman, eds.), pp. 248–254, Academic Press, New York.

Meyerson, B., 1972, Latency between intravenous injection of progestins and the appearance of estrous behavior in estrogen-treated ovariectomized rats, *Horm. Behav.* **3**:1.

Moguilewsky, M., and Raynaud, J-P., 1979, The relevance of hypothalamic and hypophyseal progestin receptor regulation in the induction and inhibition of sexual behavior in the female rat, *Endocrinology* **105**:516.

Notides, A. G., 1970, Binding affinity and specificity of the estrogen receptor of the rat uterus and anterior pituitary, *Endocrinology* **87**:987.

O'Malley, B. W., Schrader, W. T., and Spelsberg, T. C., 1973, Receptor interactions with the genome of eucaryotic target cells, in: *Receptors for Reproductive Hormones* (B. W. O'Malley and A. Means, eds.), pp. 174–196, Plenum Press, New York.

Pavlik, E. J., and Coulson, P. B., 1976, Modulation of estrogen receptors in four different target tissues: Differential effects of estrogen vs. progesterone, *J. Steroid Biochem.* **7**:369.

Reuter, L. A., and Lisk, R. D., 1976, Progesterone may act at hypothalamus and pituitary by way of enhancement of oestrogen retention, *Nature* **262**:790.

Scatchard, G., 1949, The attractions of proteins for small molecules and ions, *Ann. N.Y. Acad. Sci.* **51**:660.

Schwartz, S. M., Blaustein, J. D., and Wade, G. N., 1979, Inhibition of estrous behavior by progesterone in rats: Role of neural estrogen and progestin receptors, *Endocrinology* **105:**1078.

Shaw, L. M. J., and Huang, R. C., 1970, A description of two procedures which avoid the use of extreme pH conditions for the resolution of components isolated from chromatins prepared from pig cerebellar and pituitary nuclei, *Biochemistry* **9:**4530.

Spelsberg, T. C., Steggles, A. W., and O'Malley, B. W., 1971, Progesterone binding components of chick oviduct, *J. Biol. Chem.* **246:**4188.

Talley, D., Li, J. J., Li, S., and Villee, C. A., 1975, Biochemical comparison of estrogen receptors of the hamster hypothalamus and uterus, *Endocrinology* **96:**1135.

Zeilmaker, G. H., 1966, The biphasic effect of progesterone on ovulation in the rat, *Acta Endocrinol.* **51:**461.

12

A Cellular Polyprotein from Bovine Hypothalamus
Structural Elucidation of the Precursor to the Nonapeptide Hormone Arginine Vasopressin

Dietmar Richter and Hartwig Schmale

1. Introduction

Although it has been recognized for a number of years that hormones may exert their effect on target tissues by acting on gene expression either directly, or indirectly, via a second messenger, relatively very little is known about the mechanisms involved. This situation has been made more complicated by the recent discoveries of the cellular polyproteins (Richter *et al.*, 1980a), where a single primary translation product gives rise to several functionally distinct biological entities by a subsequent series of specific proteolytic cleavages. The best-described example to date is opiocortin, which as a single polypeptide is a precursor to adrenocorticotropin, the endorphins, and the melanocyte-stimulating hormones. There it has been demonstrated that regulation of peptide hormone synthesis takes place not only at the transcriptional but also at the posttranslational level by differential processing and modification of the primary translation product (Herbert, 1981). Since the synthesis of peptide hormones is recognized as being

Dietmar Richter and Hartwig Schmale ● Institut für Physiologische Chemie, Abteilung Zellbiochemie, Universität Hamburg, D-2000 Hamburg 20, G.F.R..

itself under hormonal control, if only in the form of an end product negative-feedback inhibition, these polyproteins provide an important model system in which to examine regulation of gene expression (Numa and Nakanishi, 1981).

The basis for any study of regulation is a detailed knowledge of the mechanism of biosynthesis and structure of the primary translation products; therefore we have initially concentrated our research efforts on the biosynthesis of the hypothalamic nonapeptide hormones arginine vasopressin (AVP) and oxytocin (OT). Both hormones are synthesized principally in the supraoptic and paraventricular nuclei of the hypothalamus together with their corresponding "carrier" proteins, the neurophysins (Brownstein *et al.*, 1980). Within these nuclei oxytocin and arginine vasopressin appear to be synthesized in different populations of magnocellular neurones, where they are packaged into neurosecretory vesicles and axonally transported into the neurohypophysis, to be either stored in the nerve endings or secreted into the blood stream.

A number of years ago Sachs *et al.* (1969) suggested that arginine vaso-pressin and its corresponding neurophysin may be synthesized as a common precursor, which may be cleaved by specific proteolytic enzymes to give rise to the functional end products. The hypothesis was strengthened when it was found that Brattleboro rats, which have hereditary diabetes insipidus, are unable to synthesize arginine vasopressin and one species of neurophysin. Subsequently, this polyprotein model gained further support when precursors to neurophysins were isolated from the hypothalami of rats (Brownstein and Gainer, 1977) and mice (Lauber *et al.*, 1979) of approximately 20,000 and 17,000 daltons, re-spectively. Pulse–chase experiments *in vivo* showed that these polypeptides, which cross-reacted with antisera raised against neurophysin, were processed to smaller polypeptides of about the same size as neurophysin itself (approximately 10,000 daltons). Moreover, treatment of one of the precursors with trypsin gave rise to an oligopeptide which cross-reacted with arginine vasopressin antisera (Russell *et al.*, 1979), though no cross-reactivity could be detected with the original longer precursor. As *in vitro* translation systems became available, neurophysin precursors were detected in wheat germ, rabbit reticulocyte lysate (Schmale *et al.*, 1979; Giudice and Chaiken, 1979; Lin *et al.*, 1979), and *Xenopus laevis* oocyte (Schmale and Richter, 1980) systems which had been supplemented with hypothalamic mRNA. Recently, we have been able to identify, using spe-cific antisera, both arginine-vasopressin- and neurophysin-II (Np II)-like amino acid sequences within a single common precursor polypeptide synthesized by the *in vitro* translation of bovine hypothalamic mRNA, without recourse to trypsinization (Schmale and Richter, 1980; 1981a). Antisera and their use in identifying antigenic sequences in longer polypeptides are, however, notorious for the ambiguity that may arise due to problems of cross-reactivity with unrelated proteins. It was therefore essential to confirm the identification of the AVP–Np II common precursor by a means independent of antisera, and if possible to elaborate its primary structure. This chapter reports the elucidation of the intramolecular organization of the AVP–Np II precursor by tryptic fingerprint analysis and by recombinant DNA technology.

2. Structural Organization of the AVP–NP II Precursor

2.1. Immunological Identification

When bovine hypothalamic mRNA is translated in a wheat germ or a rabbit reticulocyte lysate system, a product with a mol. wt. of 21,000 (21K) reacts specifically with both anti-Np II and anti-AVP antisera (Schmale and Richter, 1981a). A second translation product with a mol. wt. of 18,000 (18K) reacts only with anti-Np II, but not with anti-AVP antisera. Its N-terminus can be labeled with N-formyl[^{35}S]methionine (Schmale and Richter, 1981b; Richter *et al.*, 1982) and its mRNA appears to comigrate on isokinetic sucrose density gradients with that for the 21K preproform (Richter *et al.*, 1980b).

On cotranslational addition of nuclease-treated dog pancreas microsomal membranes to a rabbit reticulocyte lysate system, a single major polypeptide is evident that reacts with both anti-Np II and anti-AVP antisera (Schmale and Richter, 1981a). It has an apparent mol. wt. of 23,000 daltons (23K) and is glycosylated (Schmale and Richter, 1981a; Ivell *et al.*, 1981). Such membrane preparations contain a specific peptidase activity that removes the N-terminal signal peptide as the nascent protein is synthesized and transported across the microsomal membrane. The size of this signal sequence was estimated by translating bovine hypothalamic mRNA in the presence of ascites tumor cell membranes (Schmale and Richter, 1981a) which had been pretreated *in vivo* with tunicamycin to eliminate their glycosylation capability, but leave their signal peptidase activity intact. A single product with an M_r of 19,000 (19K) was the result, which reacted with both anti-AVP and anti-Np II antisera, implying the loss of a signal sequence of about 15–20 amino acids.

The third translation system we used is the *Xenopus laevis* oocyte, where microinjected bovine hypothalamic mRNA has been shown to direct the synthesis of several polypeptides reacting with both anti-Np II and anti-AVP antisera (Schmale and Richter, 1980). Dominant among these specific translation products is a polypeptide of 23,000 daltons (23K), which is glycosylated (Ivell *et al.*, 1981) and which corresponds to the 23K proform identified in the rabbit reticulocyte lysate system. In addition, a smaller product of 14,000 daltons (14K) is synthesized in the oocyte system. Pulse–chase experiments indicate that the 14K is an intermediate product derived from the larger precursor (D. Richter and H. Schmale, unpublished). Because the 14K polypeptide shows antigenic reactivity towards both anti-AVP and anti-Np II antisera, it appears likely that this molecule is a direct precursor to the end products AVP (mol. wt. 1200) and Np II (mol. wt. 10,000), and that these may thus be close neighbors in the primary sequence.

That the 23K proform is a glycoprotein has been shown by several criteria (Ivell *et al.*, 1981). First, the 23K product is quantitatively bound to concanavalin A-Sepharose; it can be released from the lectin on subsequent washing with α-methyl mannoside. Second, when the 23K proform is treated with jackbean α-mannosidase, the mannose-rich side-chains are removed to yield a product migrating at approximately 19K, comparable to the proform obtained by translation with tunicamycin-treated membranes. Third, in the oocyte system incorporation of mannose, glucosamine, and fucose into the 23K proform has been detected.

Since neither the AVP nor Np II amino acid sequences contain a typical glycosylation site (Asn-X-Ser or Asn-X-Thr) the carbohydrate chain(s) most likely is located in the cryptic region of the precursor. Whether glycosylation plays a role in specifying posttranslational modification steps, such as exposing or masking a cleavage site remains to be elucidated. Although the bovine precursors to AVP–Np II and OT–Np I are structurally remarkably homologous molecules, at least in their noncryptic sequences (78% homology), only the AVP–Np II proform carries a carbohydrate chain implying that glycosylation is not necessarily a prerequisite for posttranslational cleavages.

2.2. Tryptic Peptide Mapping

In order to get further information on the intramolecular organization of the AVP–Np II precursor, the 21K preform, the 23K proform, and the 14K intermediate were labeled with [^{35}S]cysteine and/or [^{125}I]tyrosine and compared by tryptic peptide mapping. For this experiment bovine hypothalamic mRNA was translated either in a rabbit reticulocyte lysate system or in *Xenopus laevis* oocytes, followed by immunoprecipitation and electrophoresis on polyacrylamide/sodium dodecyl sulfate (SDS) gels. Guided by autoradiograms prepared from the gels, the radioactive bands corresponding to the precursors were excised and electroeluted. The individual proteins were concentrated by trichloroacetic acid precipitation in the presence of carrier bovine Np II, which also served as internal standard, oxidized in performic acid and digested with trypsin (Schmale and Richter, 1981c). Fingerprints were then prepared of the resulting tryptides by two-dimensional electrophoresis–chromatography on cellulose thin-layer plates, followed by ninhydrin staining and autoradiography.

As reported, tryptic peptide mapping of bovine Np II yields nine tryptides of which four contain cysteine; one of these four also yields tyrosine (Wuu and Crum, 1976). Trypsinization of AVP yields glycinamide and the fragment AVP$_{1-8}$, containing two cysteine and one tyrosine residue. As expected in all fingerprints of the 21K, 23K, and 14K [^{35}S]cysteine-labeled products, four (spots 1–4) of the nine ninhydrin-stained Np II marker tryptides coincided with the radioactive spots (Fig. 1).

When labeling with [^{125}I]tyrosine alone, only one of the Np II tryptides (spot 4) was labeled having a slightly altered mobility with respect to the authentic fragment, probably due to the iodination (Fig. 2). The additional spot (closed circle) observed in the autoradiogram (Fig. 2) has not yet been identified; it may consist of AVP and/or a tryptide of the cryptic region (see also Fig. 4).

A fifth radioactive spot was evident in the fingerprints of the 21K, 23K, and 14K [^{35}S]cysteine-labeled precursors. Whereas this spot was similar for the 23K and 14K precursors (spot 6), it migrated differently in the fingerprint of the 21K precursor (spot 5, Fig. 1). Spot 6 was identified as the tryptide of AVP, since it comigrated precisely with authentic AVP_{1-8} in two different electrophoretic systems (Schmale and Richter, 1981c). Spot 5, only observed in the 21K fingerprint, clearly did not comigrate with AVP_{1-8}; nonetheless, it apparently contained amino acid sequences immunologically identical to AVP, since it cross-reacted with anti-AVP antiserum.

These results pointed to a single unambiguous model for the structure of the AVP–Np II precursor. The key difference between the 21K and 23K or 14K precursors is that the 23K and 14K lack the signal sequence; the AVP-like spot 5 appears to be substituted by AVP_{1-8} (spot 6) when the signal peptide is removed from the 21K preform. Signal sequences are not separated from their proforms by basic amino acids and consequently not accessible to trypsin. Spot 5 would therefore represent AVP_{1-8} with an N-terminal extension comprised of part or all of the signal sequence. Since the 14K product implies the proximity of AVP and Np II within the molecule, Np II would be C-terminal to the AVP. Hence the cryptic region with the site of glycosylation should be at the C-terminus of the AVP–Np II precursor. By analogy with the primary sequences available for other peptide precursors, pairs of basic amino acids were predicted to separate AVP from Np II and the latter polypeptide from the succeeding C-terminal glycoprotein (Schmale and Richter, 1981c). A similar model was postulated for the rat AVP-Np precursor (Russell *et al.*, 1981).

2.3. Nucleotide Sequencing of Cloned cDNA

The preceding experiments provide valuable information on the hypothetical structure of the AVP–Np II polyprotein, as well as corroborating the identification by specific antibodies. However, conclusive structural evidence can only be obtained by having the complete primary amino acid sequence of the precursor. To this end double-stranded cDNA was prepared by standard procedures from total poly A$^+$ RNA isolated from bovine hypothalami and cloned in *Escherichia coli* 5K, using plasmid pBR 322 as vector (Land *et al.*, 1982). Since the proportion of total poly A$^+$ RNA estimated to code for the AVP–Np II precursor is

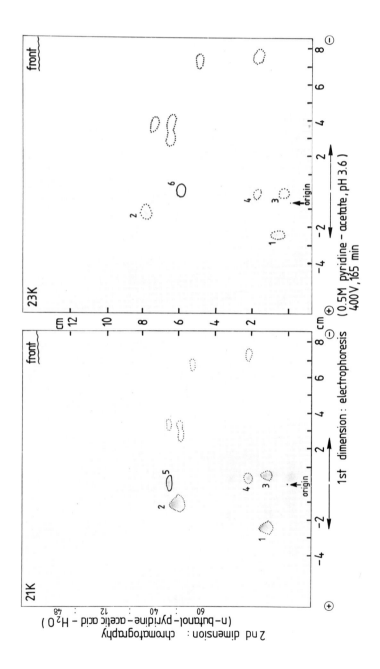

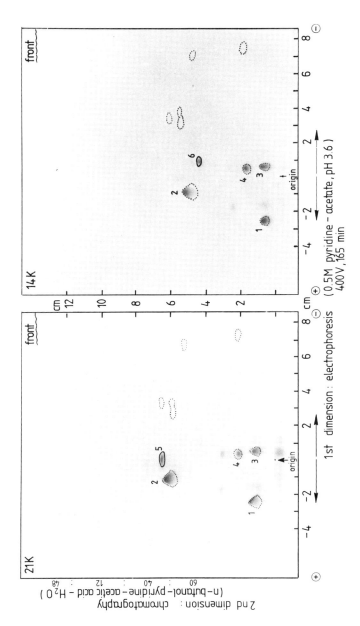

Figure 1.   Autoradiograms of fingerprints labeled *in vitro* with [^{35}S]cysteine of tryptic peptides from performic acid oxidized 21K, 23K, and 14K. After chromatography, marker tryptides from bovine Np II (dotted circles) and AVP$_{1-8}$ (spot 6, closed circles) were visualized with ninhydrin and the chromatograms were exposed to films for 1 week. Similar results were obtained in another two-dimensional system using 30% formic acid in the first dimension and n-butanol/pyridine/acetic acid/H$_2$O (60:40:12:48 v/v/v/v) in the second dimension. For experimental details see Schmale and Richter, 1981c.

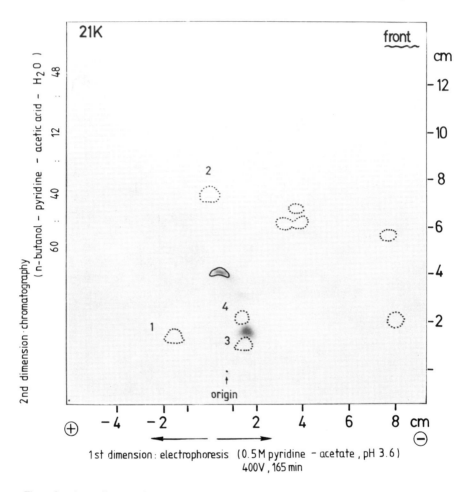

Figure 2.   Autoradiogram of a fingerprint of tryptides from the performic-acid-oxidized 21K precursor to AVP–Np II labeled by *in vitro* incubation (Schmale and Richter, 1981a) with [^{125}I]tyrosyl-tRNA (NEN). For other details see legend to Fig. 1 and Schmale and Richter, 1981c.

approximately 1% (Richter *et al.*, 1982), the clones were preliminarily screened using AVP–Np II precursor mRNA enriched 10-fold by isokinetic sucrose gradient centrifugation and labeled with ^{32}P. Of the 550 initial colonies, 55 proved positive at this first screening (Land *et al.*, 1982). However, because of the low yields of mRNA obtained from bovine hypothalami, the usual positive hybrid screening could not be applied, and an alternative screening procedure had to be found.

A strategy was adopted whereby restriction endonuclease cleavage sites in an unknown DNA sequence could be unambiguously predicted from the known sequence of amino acids. Few restriction endonucleases have recognition sites fulfilling the necessary requirements. *Fnu* 4HI (GCNGC) cuts DNA at positions corresponding to the amino acid sequences ala–ala (GCNGCN), and *Sau* 96I (GGNCC) at all positions coding for gly–pro (GGNCCN). Such gly–pro dipeptides occur twice in the amino acid sequences of the neurophysins, at positions 14–15 and 23–24 (Fig. 3). On treatment with *Sau* 96I the DNA coding for neurophysin should then be digested to yield a 27 base pair (bp) fragment. Computer analysis showed the possible existence of one extra restriction site for *Sau* 96I within the 27-bp fragment. If the site were present, the 27-bp oligonucleotide would be cleaved to yield 14-bp and 13-bp fragments. Using this system, plasmid DNAs prepared from the 55 colonies positive in the preliminary screening were divided into 11 groups of five and digested with *Sau* 96I. After 3'terminal labeling with ^{32}P using *E. coli* DNA polymerase I (Klenow fragment) and α[^{32}P]-dGTP, the digests were run on polyacrylamide gels and those plasmids from groups showing a 27-bp cleavage product were then partially sequenced (Land *et al.*, 1982). Among the eight plasmids that gave a 27-bp fragment, analysis showed three to contain Np II-specific sequences. Using one of these positive plasmids, labeled with ^{32}P, all the 550 original recombinant colonies were rescreened, to reveal altogether seven positive clones, five of which carried sequences homologous to the bovine AVP–Np II precursor, and two specific for the oxytocin–neurophysin I precursor; this cross-hybridized because of the high sequence conservation between Np I and Np II.

Plasmid pVNp II-1 was selected for further investigation because it contained the longest cDNA insert (approximately 750 bp) specific for the AVP-Np II precursor. The nucleotide sequence was determined by the Maxam and Gilbert (1977) method from restriction fragments prepared using *Sma*I, *Xma*III, *Ava*I, and *Pvu*II. Although a method was used to prepare the cDNA which preserved as much as possible of the 5' end (Land *et al.*, 1981), primer extension sequencing was additionally carried out to give as complete a picture as possible of this 5' region of the mRNA. The entire nucleotide sequence is shown in Fig. 4, together with the amino acid sequence that can be predicted from it.

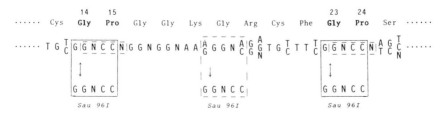

Figure 3. Cleavage sites of the restriction endonuclease *Sau* 96I predicted from the amino acid sequence of Np II. Definite sites are framed by solid lines, possible sites by broken lines.

```
5'  -100
(NGCACAGUCUACAGAGCAGCACUGGCACGUGUGCCCACG)CGUGCCAGG

                                                            -50
AUG CCC GAC ACA CUG CCC GCC UGC UUC CUC AGC  CUG CUG GCC UUC ACC UCU GCU UGC
MET PRO ASP THR LEU PRO ALA CYS PHE LEU SER  LEU LEU ALA PHE THR SER ALA CYS
-19                                                                      -1

                                                    50
 UAC UUC CAG AAC UGC CCA AGG GGC  GGC AAG AGG  GCC AUG UCC GAC CUG GAG
 TYR PHE GLN ASN CYS PRO ARG GLY  GLY LYS ARG  ALA MET SER ASP LEU GLU
  1                          10

                      100                                            150
CUG AGA CAG UGU UGC GAG GGC GGG CCC UGC GGG AAA GGG GGC GGG CUG UGC UGC GGC
LEU ARG GLN CYS CYS GLU GLY GLY PRO CYS GLY LYS GLY GLY GLY LEU CYS CYS GLY
         20                                                    40

                               200
CUG GGC UUC CAG AAC UAC CUG CCC UCG CCC UGC CAG GAC
LEU GLY PHE GLN ASN TYR LEU PRO SER PRO CYS GLN ASP
                    60

                         250
GGC CAG AAG CCC UGC GGC AGC GGG GGU CGG UGC GCC GCC GCU GCA GAC GAU AAC AGC
GLY GLN LYS PRO CYS GLY SER GLY GLY ARG CYS ALA ALA ALA ALA ASP ASP ASN SER
                    80

                        300
GAG CCC GAG UGC CGC GUC CGC GUU CGC GCC AGC AAC GCG ACC GCG GUG
GLU PRO GLU CYS ARG VAL ARG VAL ARG ALA SER ASN ALA THR ALA VAL
100

                              350
GAC GGG CCC AGC GGG GCC CUG CUG CGG CUG GUG CAG CUG CUG GUG CAG CUG CUG CUG
ASP GLY PRO SER GLY ALA LEU LEU ARG LEU VAL GLN LEU LEU VAL GLN LEU LEU LEU
                              120

                            400
GAC GGG CCG GCG CUG GGG CCG GAG GCC CCG GCG GAG CUG GCG CCC GCC CAG
ASP GLY PRO ALA LEU GLY PRO GLU ALA PRO ALA GLU LEU ALA PRO ALA GLN
140

CCC GGC GUC UAC UGA
PRO GLY VAL TYR ***
147

                     450
GGGCGCCCCCCCUCCCCACCCCUGCCCUCGCAGCACGAAAAUAAACGUUUUAAAGGC(A)~150  3'
                                         500
```

Figure 4. Nucleotide sequence of AVP–Np II precursor mRNA from bovine hypothalamus and the deduced amino acid sequence. The nucleotide numbers are shown above the sequence. Nucleotide residues are numbered in the direction 5'→3' in the mRNA strand, beginning with the first residue in the coding region for arginine vasopressin. The nucleotides upstream from residue 1 are indicated by negative numbers. The predicted amino acid sequence is numbered by designating as 1 the first residue (Cys) of arginine vasopressin. The amino acids constituting the putative signal peptide are indicated by negative numbers, those towards the carboxyl terminus of the precursor have positive numbers. The untranslated sequence at the 5' terminus, shown in parenthesis, is derived from primer extension sequencing of the mRNA and has not been confirmed by sequencing of cloned DNA.

3. Primary Structure of the AVP–Np II Common Precursor

The amino acid sequence of the AVP/Np II precursor (schematically represented in Fig. 5) substantiated fully our deductions based on the tryptic fingerprint analysis. There appears to be a signal sequence of 19 amino acids, directly followed by the sequence of AVP, without an intervening basic amino acid. Arginine vasopressin is in turn followed by the triplet gly–lys–arg, and the bovine Np II sequence. The latter is then separated by a single arginine from a polypeptide of 39 amino acids, which contains a single glycosylation site Asn-X-Thr at residue 114–116, and then a stop codon. This glycoprotein, located at the C-terminus of the AVP common precursor, is completely homologous with

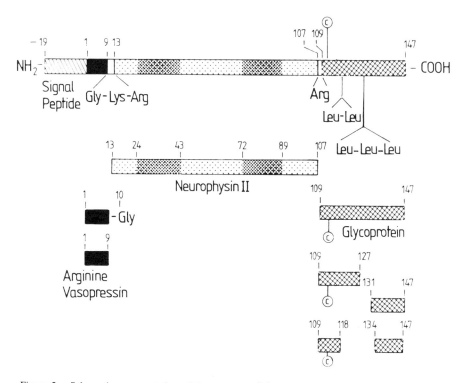

Figure 5. Schematic representation of the structure of the common precursor to AVP and Np II from bovine hypothalamus. Numbering of the amino acid residues is as in Fig. 4. The putative signal peptide (−19 to −1) is represented by the hatched bar. Vasopressin (closed bar) extends from amino acid residue 1 to 9. Neurophysin II (13–107), shown as a shaded bar, exhibits areas of internal homology (dark shading) located between amino acid residues 24 and 43 and 72 and 89. The glycoprotein part of the precursor (109–147, cross-hatched bar) carrying a carbohydrate chain (C) at position 114 is further cleaved into several peptides of 10(109–118), 18(109–127), 17(131–147), and 14(134–147) amino acid residues.

a protein isolated and sequenced from the pituitaries of several species (Holwerda, 1972; Smyth and Massey, 1979; Seidah *et al.*, 1981), and whose glycosylation at the Asn-X-Thr site has a similar sugar composition to that estimated for the 23K proform synthesized in frog oocytes (Ivell *et al.*, 1981). Nothing is known at all about its biological function or its localization, though several fragments have been isolated from pituitary extracts (Smyth and Massey, 1979). It would appear to be well-conserved during evolution, implying some valuable function to the organism.

4. Glycine as Signal for Amidation of Oligopeptides

A number of peptide hormones have been found to contain an amide group at their C-terminus, which, for instance, may serve to increase receptor specificity. The biosynthetic origin of this amide group has been obscure. However, with the elucidation of the melittin precursor sequence (Suchanek and Kreil, 1977) attention was drawn to the possible role of a glycine residue which immediately follows C-terminally the biological peptide within the precursor sequence. Since

Table 1. Partial Amino Acid Sequence of Precursors to
Oligo-peptides Amidated at the Carboxy Terminus

Melittin[1]	-Arg-Lys-Arg-Gln-Gln-Gly-COOH
α-MSH[2]	-Trp-Gly-Lys-Pro-Val-Gly-Lys-Lys-
γ-MSH[3]	-Arg-Trp-Asp-Arg-Phe-Gly-Arg-Arg-
Vasopressin[4]	-Asn-Cys-Pro-Arg-Gly-Gly-Lys-Arg-
Oxytocin[5]	-Asn-Cys-Pro-Leu-Gly-Gly-Lys-Arg-
Caerulein[a]	-Gly-Trp-Met-Asp-Phe-Gly-Arg-Arg-
Gastrin[6]	-Gly-Trp-Met-Asp-Phe-Gly-Arg-Arg-
Calcitonin[7]	-Gly-Val-Gly-Ala-Pro-Gly-Lys-Lys-
Human joining peptide[8]	-Pro-Gly-Pro-Arg-Glu-Gly-Lys-Arg-
Egglaying hormone[9]	-Arg-Leu-Leu-Glu-Lys-Gly-Lys-Arg-

[1] Suchanek and Kreil (1977).
[2] Nakanishi *et al.* (1979).
[3] Shibasaki *et al.* (1980).
[4] Land *et al.* (1982).
[5] Land *et al.*, in press.
[6] Yoo *et al.* (1982).
[7] Amara *et al.* (1980).
[8] Seidah *et al.* (1980).
[9] Scheller *et al.* (1982).
[a] Hoffmann, W. and Kreil, G., personal communication.

then, several precursors to amidated peptide hormones have been sequenced, including now the AVP–Np II precursor. In all cases (Table I) there is a glycine C-terminal to the active peptide sequence, where this occurs in a longer precursor sequence. The glycine is then followed by a pair of basic amino acids, known to be a typical prohormone processing signal. The enzyme(s) that may be involved in this amidation reaction are currently being characterized by Smyth's group (D. Smyth, personal communication). It appears that prior to amidation the cleavage of the adjacent basic amino acids has to occur.

5. The 18K Precursor

One still-unresolved question concerns the nature of the 18K polypeptide synthesized *in vitro*. Our earlier speculations that it may be the product of an early termination event (Richter *et al.*, 1980a,b) seem unlikely now that we have the complete primary sequence. The tryptic fingerprint analysis confirms that the cysteine-containing tryptic peptides of bovine Np II are all accurately present, and that, as predicted by the immunological evidence, no tryptic peptide can be identified corresponding to AVP (Fig. 6).

However, it is not possible to determine whether the 18K is the product of an independent mRNA, akin to the neurophysin III found in rats (Brownstein *et al.*, 1980), or whether it is an artifact of the *in vitro* translation system. For instance, due to less stringent constraints, a polypeptide may be synthesized from the mRNA, but initiated at a later position (aa 33 of the preproform), thus eliminating the arginine vasopressin sequence.

Examples for such a second initiation site within a given mRNA are known only from viral polyproteins. According to the hypothesis of Kozak (1978), ribosomes bind at the cap region of the 5' noncodon region and scan the mRNA until they "detect" the first AUG initiation codon. In the case of the 18K precursor, the 5' noncodon region and the 96 nucleotides of the translatable region may well be part of a 5'-terminal double-stranded RNA stem suppressing initiation at the first AUG codon (position −43). If this is the case, the 18K product should lack 32 amino acids of the 21K preproform. In preliminary experiments ^{35}S-cysteine-labeled 21K and 18K products were isolated from polyacrylamide/SDS gels and subjected to cleavage by cyanogen bromide. According to the amino acid sequence of the 21K precursor (Fig. 4), cyanogen bromide cleavage yields two ^{35}S-cysteine-labeled products, one with a mol. wt. of about 3000 and the other of 18,000 (Fig. 7). The latter comigrates with the *in vitro* synthesized 18K product, which is not altered by the cyanogen bromide cleavage. Further analyses are required to confirm the proposed second initiation site of the AVP–Np II precursor which would represent another parallel to viral polyproteins.

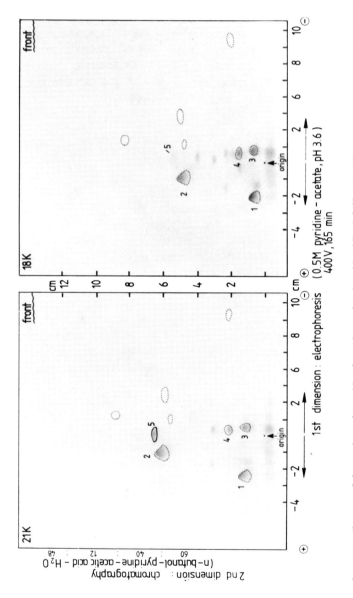

Figure 6. Autoradiogram of fingerprints of tryptides from the performic-acid-oxidized 21K and 18K precursor labeled *in vitro* with [^{35}S]cysteine. For other details see legend to Fig. 1.

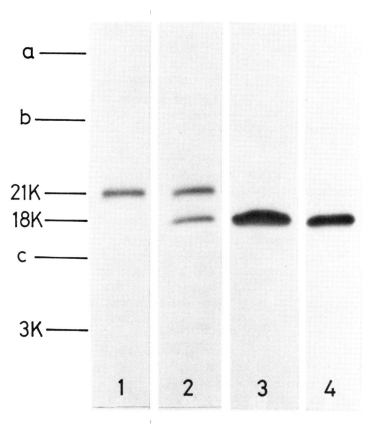

Figure 7. Autoradiogram of the 21K and 18K products analyzed on SDS/polyacrylamide gels. Poly A$^+$ RNA from bovine hypothalamus was translated in the presence of [^{35}S]cysteine, immunopre-cipitated with anti Np II-antiserum, and analyzed by SDS/15% polyacrylamide gel electrophoresis (Schmale and Richter, 1981a). The 21K and 18K products were electroeluted, subjected to cyanogen bromide cleavage (Chen-Kiang *et al.,* 1979), and electrophoresed again. The dried gels were au-toradiographed. Lane 1 and 4 represent the untreated 21K and 18K products, respectively; lanes 2 (21K) and 3 (18K) after cyanogen bromide treatment. The 21K product treated by cyanogen bromide yielded three products: 21K, 18K, and 3K (lane 2), apparently the cleavage was not complete leaving some 21K material intact. a, b, and c are the marker proteins ovalbumin, 46,000; carbonic anhydrase, 30,000; lysozyme 14,300.

6. Concluding Remarks

We now have a picture of the first steps in the pathway of biosynthesis leading to the release of the nonapeptide hormone. A number of processing steps are clearly involved, probably coordinated to physical transport and packaging of the molecule in the neurosecretory magnocellular neurones. It will be extremely interesting to find out how the nascent polypeptide is folded, to allow for subsequent processing and to provide the confirmation of neurophysin which is an integral part of its function as a carrier of the nonapeptide. It also remains completely open as to what function is served during these processing and transport stages, or subsequently, by the C-terminal glycopolypeptide, which it should be noted is completely absent from the oxytocin precursor.

The structure of the AVP-Np II precursor indicates that this polyprotein has interesting features. Firstly, like other polyproteins it is composed of several distinct biological entities. The processing of these polyproteins is generated by proteolytic enzymes, a reaction often directed by pairs of basic amino acids. The maturation process can be accompanied by modifications of the biologically active peptide, e.g. amidation of the arginine vasopressin. This amidation reaction requires a glycine residue to the hormone and is found in several precursors to amidated peptides. Secondly, neurophysin II and the glycoprotein are separated by only one arginine residue (another example would be the human tumor somatostatin precursor where somatostatin-28 is also separated from the cryptic region by one basic amino acid residue). Thirdly, it is probable that all three components of the vasopressin precursor are functional units, including the glycoprotein; although a physiological role for the latter moiety has not yet been identified, its high sequence conservation between species would imply some function. Thus, besides processing recognition signals, the vasopressin precursor does not include long "spacer" sequences separating functional units as found in the opiocortin and enkephalin precursors.

The primary sequence information reported here offers a framework for studying the regulation of the expression of the vasopressin gene at various levels. The Brattleboro rat model should be extremely helpful in answering how regulation of such composite preprohormones can be monitored at the level of transcription, translation and posttranslation. Rats not only offer a well characterized *in vivo* system but have the great advantage that in the Brattleboro rat, a mutant exists with hereditary hypothalamic diabetes insipidus. Studies have shown that these animals lack biologically active vasopressin as well as its corresponding neurophysin carrier (Valtin *et al.,* 1974). Since cDNA probes encoding the bovine AVP-Np II precursor will hybridize to rat hypothalamic mRNA and to the rat AVP-Np gene, tools are now available to study vasopressin gene expression in Brattleboro rats. Preliminary experiments analyzing supraoptic nuclei mRNA from Brattleboro rats either in cell-free translation experiments or by hybridization suggest that these animals lack AVP-Np mRNA. Exact confirmation, however will have to wait until DNA probes are available, encoding sequences uniquely present in the AVP-Np precursor and not occurring in the partially homologous OT-Np precursor.

ACKNOWLEDGMENTS. We thank Dr. Richard Ivell for stimulating discussions during preparation of the manuscript; S. Heinsohn, M. Hillers, A. Ladak, and

C. Schmidt were involved in various steps of the project; their help is gratefully acknowledged.

REFERENCES

Brownstein, M. J., and Gainer, H., 1977, Neurophysin biosynthesis in normal rats and in rats with hereditary diabetes insipidus, *Proc. Natl. Acad. Sci. USA* **74**:4046.

Brownstein, M. J., Russell, J. T., and Gainer, H., 1980, Synthesis, transport, and release of posterior pituitary hormones, *Science* **207**:373.

Chen-Kiang, S., Stein, S., and Udenfriend, S., 1979, Gel electrophoresis of fluorescent labeled cyanogen bromid cleavage products at the submicrogram level, *Anal. Biochem.* **95**:122.

Guidice, L. C., and Chaiken, I. M., 1979, Immunological and chemical identification of a neurophysin-containing protein coded by messenger RNA from bovine hypothalamus, *Proc. Natl. Acad. Sci. USA* **76**:3800.

Herbert, E., 1981, Discovery of pro-opiomelanocortin—a cellular polyprotein, *Trends Biochem. Sci.* **6**:184.

Holwerda, D. A., 1972, A glycopeptide from the posterior lobe of pig pituitaries, *Eur. J. Biochem.* **28**:340.

Ivell, R., Schmale, H., and Richter, D., 1981, Glycosylation of the arginine vasopressin/neurophysin II common precursor, *Biochem. Biophys. Res. Commun.* **102**:1230.

Kozak, M., 1978, How do eukaryotic ribosomes select initiation regions in messenger RNA? *Cell* **15**:1109.

Land, H., Grez, M., Hauser, M., Lindemaier, W., and Schütz, G., 1981, 5′ Terminal sequences of eukaryotic mRNA can be cloned with high efficiency. *Nucleic Acid Res.* **9**:2251.

Land, H. Schütz, G., Schmale, H., and Richter, D., 1982, Nucleotide sequence of cloned DNA encoding the bovine arginine vasopressin–neurophysin II precursor, *Nature* **295**:299.

Lauber, M., Camier, M., and Cohen, P., 1979, Immunological and biochemical characterization of distinct high molecular weight forms of neurophysin and somatostatin in mouse hypothalamus extracts, *FEBS Letts.* **97**:343.

Lin, C., Joseph-Bravo, P., Sherman, T., Chan, L., and McKelvy, J. F., 1979, Cell-free synthesis of putative neurophysin precursors from rat and mouse hypothalamic poly(A)-RNA, *Biochem. Biophys. Res. Commun.* **89**:943.

Maxam, A., and Gilbert, W., 1977, A new method for sequencing DNA, *Proc. Natl. Acad. Sci. USA* **74**:560.

Nakanishi, S., Inoúe, A., Kita, T., Nakamura, M., Chang, A. C. Y., Cohen, S. N., and Numa, S., 1979, Nucleotide sequence of cloned cDNA for bovine corticotropin-β-lipotropin precursor, *Nature* **278**:423.

Numa, S., and Nakanishi, S., 1981, Corticotropin-β-lipotropin precursor—a multi-hormone precursor—and its gene, *Trends Biochem. Sci.* **6**:274.

Richter, D., Ivell, R., and Schmale, H., 1980a, Neuropolypeptides illustrate a new perspective in mammalian protein synthesis—the composite common precursor, in: *Biosynthesis, Modification, and Processing of Cellular and Viral Polyproteins* (G. Koch and D. Richter, eds.), pp. 5–13, Academic Press, New York.

Richter, D., Schmale, H., Ivell, R., and Schmidt, C., 1980b, Hypothalamic mRNA-directed synthesis of neuropolypeptides: Immunological identification of precursors to neurophysin II/arginine vasopressin and to neurophysin I/oxytocin, in: *Biosynthesis, Modification, and Processing of Cellular and Viral Polyproteins* (G. Koch and D. Richter, eds.), pp. 43–66, Academic Press, New York.

Richter, D., Schmale, H., Ivell, R., and Kalthoff, H., 1982, Cellular polyproteins from pituitary and hypothalamus: Composite precursors to oligopeptide hormones, in: *Hormonally Active Brain*

Peptides: Structure and Function (K. W. McKerns and V. Pantić, eds.), pp. 581–598, Plenum Press, New York.

Russell, J. T., Brownstein, M. J., and Gainer, H., 1979, Trypsin liberates an arginine vasopressin-like peptide and neurophysin from a M_r 20,000 putative common precursor, Proc. Natl. Acad. Sci. USA 76:6086.

Russell, J. T., Brownstein, M. J., and Gainer, H., 1981, Biosynthesis of neurohypophyseal polypeptides: The order of peptide components in pro-pressophysin and pro-oxyphysin, Neuropeptides 2:59–65.

Sachs, H., Fawcett, P., Takabatake, Y., and Portanova, R., 1969, Biosynthesis and release of vasopressin and neurophysin, Recent Prog. Horm. Res. 25:447.

Schmale, H., and Richter, D., 1980, In vitro biosynthesis and processing of composite common precursors containing amino acid sequences identified immunologically as neurophysin I/oxytocin and neurophysin II/arginine vasopressin, FEBS Letts. 121:358.

Schmale, H., and Richter, D., 1981a, Immunological identification of a common precursor to arginine vasopressin and neurophysin II synthesized by in vitro translation of bovine hypothalamic mRNA, Proc. Natl. Acad. Sci. USA 78:766.

Schmale, H., and Richter, D., 1981b, RNA-directed synthesis of a common precursor to the nonapeptide arginine vasopressin and neurophysin II: Immunological identification and tryptic peptide mapping, Hoppe Seylers Z. Physiol. Chem. 362:1551.

Schmale, H., and Richter, D., 1981c, Tryptic release of authentic arginine vasopressin$_{1-8}$ from a composite arginine vasopressin/neurophysin II precursor, Neuropeptides 2:47.

Schmale, H., Leipold, B., and Richter, D., 1979, Cell-free translation of bovine hypothalamic mRNA. Synthesis and processing of the prepro-neurophysin I and II, FEBS Letts. 108:311.

Seidah, N. G., Rochemont, J., Hamelin, J., Banjannet, S., and Chrétien, M., 1981, The missing fragment of the pro-opiomelanocortin sequence and evidence for C-terminal amidation, Biochem. Biophys. Res. Commun. 102:710.

Smyth, D. G., and Massey, D. E., 1979, A new glycopeptide in pig, ox, and sheep pituitary, Biochem. Biophys. Res. Commun. 87:1006.

Suchanek, G., and Kreil, G., 1977, Translation of Melittin messenger RNA in vitro yields a product terminating with glutaminylglycine rather than with glutaminamide, Proc. Natl. Acad. Sci. USA 74:975.

Valtin, H., Stewart, J., and Sokol, H. W., 1974, Genetic control of the production of posterior pituitary principles, in: Handbook of Physiology Vol. 7, p. 737, Williams & Wilkens, Baltimore.

Wuu, T. C., and Crum, S. E., 1976, Amino acid sequence of bovine neurophysin II, Biochem. Biophys. Res. Commun. 68:634.

Index